THE BATTLE FOR
ALABAMA'S WILDERNESS

THE BATTLE FOR
ALABAMA'S WILDERNESS
Saving the Great Gymnasiums of Nature

JOHN N. RANDOLPH

The University of Alabama Press • *Tuscaloosa*

Copyright © 2005
The University of Alabama Press
Tuscaloosa, Alabama 35487-0380
All rights reserved

Designer: Michele Myatt Quinn
Typeface: AGaramond

Articles from the Associated Press reprinted with permission.

All materials quoted from the *Birmingham News* are copyright the *Birmingham News*. All rights reserved. Reprinted with permission.

All materials quoted from the *Mobile Press-Register* are copyright the *Mobile Press-Register*. All rights reserved. Reprinted with permission.

Library of Congress Cataloging-in-Publication Data

Randolph, John N. (John Nevitt), 1944–
  The battle for Alabama's wilderness : saving the great gymnasiums of nature / John N. Randolph.
     p. cm.
"Fire Ant books."
  Includes bibliographical references (p.) and index.
  ISBN 0-8173-5159-0 (pbk. : alk. paper)
  1. Wilderness areas—Alabama. 2. Nature conservation—Political aspects—Alabama. 3. Alabama—Environmental conditions. I. Title.
  QH76.5.A2R36 2005
  333.78'216'09761—dc22

2004015749

Contents

List of Maps   vii
Introduction   ix
Author's Note   xiii

PART ONE   The Sipsey Wilderness Area

1 Blanche and Mary   3
2 It's Impossible   8
3 Purity   16
4 A Seething Mass of Happenings   26
5 A Truly National System   35
6 Dedication for the Future   45

PART TWO   The Cheaha Wilderness Area

7 Nobody Here Felt Like It Was Hurting Anything   51
8 Son of RARE   57
9 Are You Lying or Ignorant?   63
10 Snail Darter for the Scenic Drive   68
11 Breakthrough   76
12 Deja-Breakthrough All Over Again   82

PART THREE   The Sipsey Wilderness Expansion

13 Perceptions   91
14 Priceless Gift of God   97
15 SWUFFL   103
16 Gutless Sellout to the Timber Industry   120
17 Elitist Birdwatchers and Backpackers   130
18 Nobody Loves You but Your Dog   145
19 Stay Out of My Business   152
20 Might As Well Move   164
21 It Cannot Be   171
22 Peace Is at Hand   179
23 In Retrospect   189

PART FOUR The Dugger Mountain Wilderness Area

24 Fire on the Mountain 195
25 Songs of the Summit 203
26 We Did This One Right 210
Epilogue 221

Notes 231
Index 249

Photographs follow page 112

Maps

Alabama National Forest Units and Their Wilderness Areas   viii

Original Sipsey Wilderness Area, Established 1975   2

Talladega Mountains and the Odum Scout Trail, Ca. 1976   50

Sipsey Wilderness Expansion as Proposed by Rep. Ronnie G. Flippo, 1982   90

Expanded Sipsey Wilderness Area and West Fork Sipsey National Wild and Scenic River, Established 1988   191

Dugger Mountain Wilderness Area, Established 1999   194

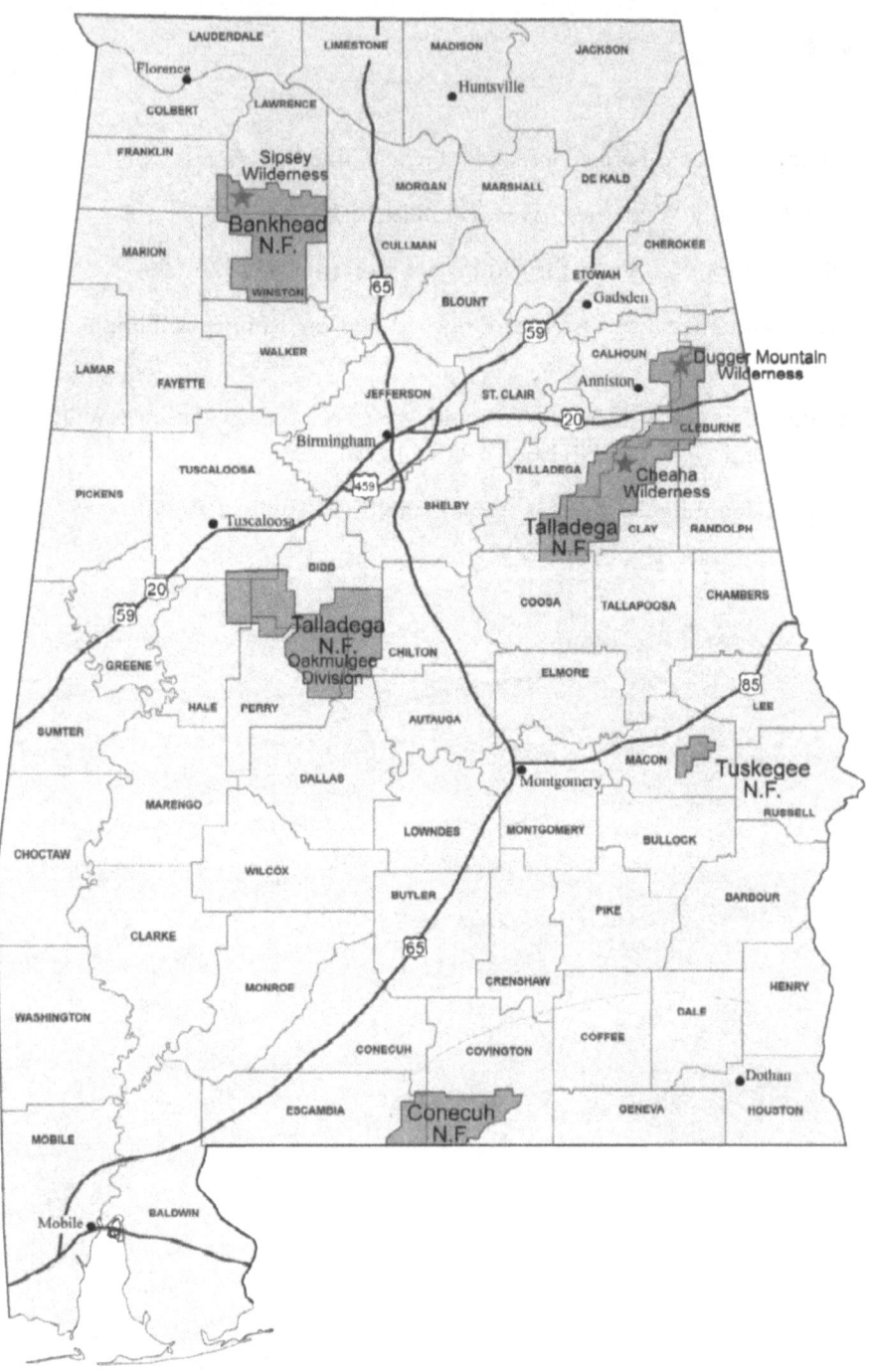

Alabama National Forest Units and Their Wilderness Areas

Introduction

I LEARNED TO SWIM in the Cahaba River.

In the early 1950s Birmingham was just large enough for children like me growing up in its suburban confines to be considered "city boys," yet the town lay in the midst of a surprising outback of ancient mountain ridges and rocky glens, seemingly inaccessible except in scattered places where farming, mining, and lumbering made inroads through the encircling forests. Driving only a few miles south from town, one quickly passed from the openness of small farms to the enveloping green of an Alabama mountain backcountry drained by the Cahaba River.

At a friend's property straddling the Cahaba, I first experienced the joy of river mud oozing between toes as I pushed myself into the center of the current, dog-paddled the pools, and slid down riffles and shoals of smooth river rock that sheltered the mysteries of free-flowing water: the mussel, the backward-darting crawfish, and the water lilies of spring. And from the first moment I touched the river, I associated the experience with freedom.

Never mind that there were always a couple of mamas around to see that we did no harm to ourselves or others; shoving off that bank into what I perceived as wilderness enveloped me in a sense of freedom. I still feel hints of that youthful simplicity, but now maturity allows me to articulate its components: the freedom to be in a wild place that is part of my heritage as an Alabamian and an American, the freedom to see a natural world and its forces at work, the wildlife it sustains, and the freedom of those forces and that world to exist.

Later, when I became a wilderness activist, I confronted the reality of what is meant by our constitutional rights to petition the government for redress and to exercise freedom of speech and of the press. In order to preserve the right to be free in wilderness, I had to learn to employ the freedom of our American system. In all meaningful ways, then, to me, wilderness *is* freedom.

The wildernesses that this book examines are the wild places that have been designated by an act of Congress and signed into law by the president of the United States as units of the National Wilderness Preservation System. Today, Alabama has three such places: the Sipsey Wilderness, Alabama's premiere natural area, and the Cheaha and Dugger Mountain Wildernesses, remnants of the final thrust of the Piedmont and the Appalachian to the coastal plain.

There was nothing easy about obtaining protection for any of these areas.

All of them required years of effort, and the Sipsey Wilderness required two separate, hard-fought campaigns to preserve its current 25,000 acres. All these battles originated at the grass roots and were led by the people you will meet in these pages.

The mother of Alabama wilderness is Mary Ivy Burks, who led the effort to establish the Sipsey Wilderness during the 1960s and '70s, and, in doing so, gave birth to the state's serious environmental movement. When asked what inspired her to enter the world of wilderness preservation, she says, simply, "passion." In my experience, that well describes most people who have immersed themselves in battles for wildness. One purpose of this book is to introduce you to Alabama's version of such folks.

The campaigns could not have been conducted, and the leaders could not have led, without the existence and efficacy of various organizations that participated in the wilderness campaigns. But it is people who lead organizations, people who have the passion to save the wilderness, and it is those individuals to whom I pay the most attention.

The effort to establish the original Sipsey Wilderness Area began as a story of ladies in tennis shoes laying siege to the U.S. Forest Service to protect a key part of the watershed of the West Fork Sipsey River. But it quickly donned the mantle of history, becoming a catalyst for national policy in the making. Its advocates faced a sometimes bewildering competition among politicians, bureaucracies, and national conservation groups, all pulling this way and that. In 1972 alone, the Sipsey was the centerpiece of no fewer than six pieces of congressional legislation.

Dugger Mountain, though it may have been the "easiest" of the campaigns, in the sense that it attracted no organized opposition, nevertheless has its own tales of frustration and intrigue, its own set of heroes.

But the main focus of this book is on the Cheaha wilderness and the Sipsey expansion efforts. These were my personal projects, and I not only retain all my files from that period but also still feel the compulsion of the time, and the stings and thrills of events. Both fights were classic, hostile confrontations with entrenched bureaucrats and intransigent politicians. Sen. Howell T. Heflin, Democrat of Tuscumbia, the Sipsey enlargement's most powerful foe, and, in the end, perhaps its only foe, plays a prominent part in my story.

I am prejudiced, and I am not shy in my advocacy of wilderness preservation, but to the extent that I am able, I let the people involved in each of the campaigns, both for and against, speak for themselves.

I also let the individual wilderness areas speak for themselves. I am not a naturalist, and this is no nature guide, no trail book. I am an advocate, and I want to share with you the people and processes involved in conducting grassroots campaigns on behalf of wilderness preservation. In my view, the value of the Sipsey, Cheaha, and Dugger Mountain Wilderness Areas is self-evident. But don't take my word for it. Go there and see for yourself.

Author's Note

MY JOB IN RECOUNTING the sagas that follow was made easier by the fact that most of the players expressed themselves on the record, usually in newspaper articles and congressional testimony, and more recently, in personal interviews. These instances are specifically footnoted. In all other cases, quotations have been reconstructed from my notes and files and my personal recollection of events.

As each tale unfolds, the reader may find the precise acreage in the various wilderness proposals and designations to be an elusive proposition. The reasons for this are many.

All the preservation campaigns began at the grass-roots level, and citizens included with their proposals their best estimates of acreage. When the U.S. Forest Service—the federal agency that manages the national forests—became involved, it would produce an initial estimate of its own, which it would "tweak" from time to time—even after legislation was passed—as a better understanding of boundaries was reached. During the RARE II process of the 1970s, the Forest Service often used both "gross" and "net" acreage figures—the former including private land on the periphery but within encircling roads, and the latter containing public lands only. Depending upon the source, press reports of the time might employ the citizens' estimates, the Forest Service's first estimates or their later refinements, or the gross or net RARE II acreages—or in some cases attributable to politicians or their staffs, figures that were simply erroneous or inexplicable. Moreover, everyone involved often found it easier to simply round off acreage figures, a practice that I have adopted.

For the purposes of this book, suffice it to say that, in round figures, the Sipsey is 25,000 acres; Dugger Mountain, 9,200 acres; and Cheaha, 7,500 acres. The narrative, I hope, will clarify the purpose and impact of significant shifts in acreage as it goes along. My advice to the reader is not to be distracted by trying to reconcile the numbers that were quoted, but to rejoice that we have the protected acreage that we do.

THE BATTLE FOR
ALABAMA'S WILDERNESS

PART ONE

The Sipsey Wilderness Area

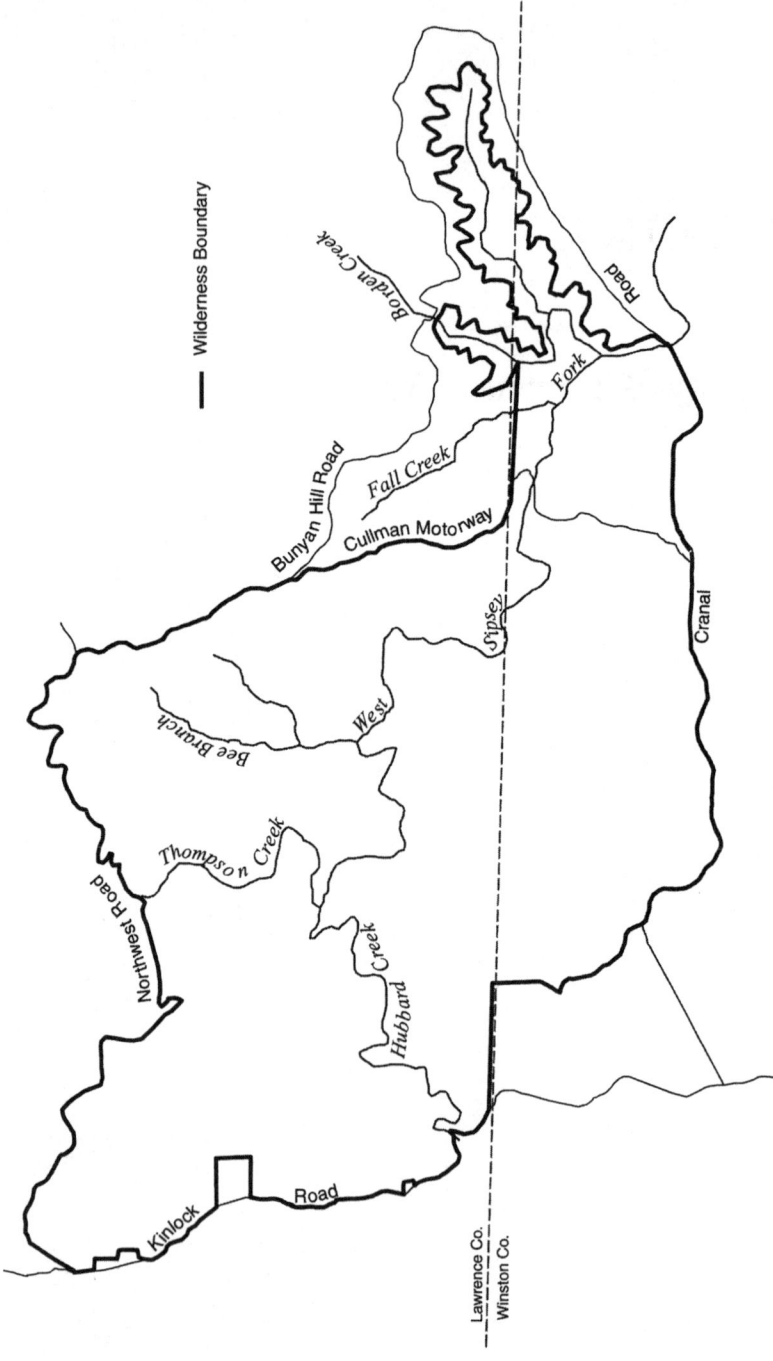

Original Sipsey Wilderness Area, Established 1975

CHAPTER ONE

Blanche and Mary

THE SIPSEY WILDERNESS AREA has influenced many lives, mine included. I have a personal interest in the Sipsey not only because I helped protect it but also because it taught me that I was something more than I thought I was. I suppose it is that way with any magnificent natural area and the people who love it.

Those whose wilderness experience is limited to the exhilaration of the mountain top might first sense an oppression within the depths of the Sipsey Wilderness, with its looming rock walls; its feel and smell of thick, damp, green gloom; and the pervasiveness of its water, evaporating, condensing, sweating, seeping, trickling, falling, running everywhere. But the intimacy of the place forces an awareness of an elemental aspect of nature's possibilities. Where the grandeur of the Rocky Mountain Front might exhibit the power and magnitude of creation, a descent into the Sipsey Wilderness reveals its subtlety and its nurturing. But it can be deceptively challenging, and dangerous. Anyone making the effort to clamber into its boulder-strewn recesses in search of the hidden waterfall will quickly draw upon reserves of stamina and dexterity similar to those required in the bigger wild: the Sipsey is damned good practice for hiking Glacier National Park. The Sipsey taught me that, beginning even as late as middle age, a person can withstand the rigors of exploration on foot and be led to undertake multiple adventures in the Grand Canyon, the Rocky Mountains, and the redrock splendor of southern Utah.

But I always come back to the Sipsey, to special places that I know as Lucy's Ditch, Tick Hollow, and tucked at the very heart of the 25,000-acre Wilderness, Back of Time Falls in Early Times Canyon—places that are mine, places that have inspired me through the years, that remind me of where I've been, what I've become.

The burden, and the honor, of preserving these places began with an earlier generation, some of whom have died, taking their stories with them. Most of the written records relating to the establishment of the original 12,700-acre Sipsey Wilderness Area in 1975 are the property of Mary and Bob Burks, cofounders of the Alabama Conservancy. I have seen no more eloquent an advocacy, and validation, for the preservation of the Sipsey than Mary Burks's testimonies before congressional committees:

> [In the canyons of the Sipsey,] abundant water and sheltering walls enclose an island of the past. Sheer sandstone cliffs, crowned with mountain laurel and filled with rare ferns and wildflowers rear 100 feet above the steep valley slopes. Hundreds of waterfalls cascade down the rock walls or plunge to the valley floor. Numerous rock shelters occur throughout the approximately 30 miles of gorges, and monstrous boulders lie shattered at the bases of the great cliffs.
>
> The Bankhead is a microcosm of Alabama plant diversity because three major land divisions meet within it. The forest is located at the southern terminus of the Cumberland Plateau where it meets the Coastal Plains. The cool summer temperatures, deep shade and constant supply of moisture in the Sipsey gorges provide a southern refuge for a relic Pleistocene Flora.... [T]he proposed Sipsey Wilderness abounds in rare plants of unusual size, diversity, distribution or rarity.... [Twenty-four] plants reach their southern limits, including the Canada Hemlock and the Sweet Birch. Twelve wildflowers on the rare and endangered species list of the Forest Service grow in the proposal area.
>
> The Sipsey has a full range of mammals—not only game, such as the abundant deer and squirrel, but otter, mink, beaver, raccoon, opossum, small rodents, shrews, bats and perhaps even a mountain lion or two. We hope two, preferably of different sexes.
>
> The Sipsey Wilderness provides a winter haven for many northern birds, shelters many additional species during migration, and attracts an unusual breeding population reaching its northern and southern nesting limits—such as the northern Whip-poor-will and the southern Chuck-will's-widow, the Scarlet Tanager and the Summer Tanager.... The gorges and uplands provide habitat for 80 species of woodland birds, of which 48 must have hardwoods to survive.
>
> Wilderness saves the flora and fauna which we do not yet have the knowledge to understand or the wisdom to appreciate. These gene pools may carry the inheritance which we must have in the future to breed survival values back into our soils, plants and animals. To destroy this area would be like burning a library before any of its books were read or even catalogued.
>
> In 1913, when ... the proposed wilderness was purchased by the United States Forest Service, its steepness, ruggedness and inaccessibility combined to give it little monetary value.... The very factors that made the Sipsey almost "worthless" in the

early days of America now give it immeasurable value. In fact, in terms of man hours of work and expenditure of personal funds by The Alabama Conservancy membership, the American people may already have spent more to protect the Sipsey area than the Forest Service paid for the land in the first place.

Walking into the deep recesses of these undisturbed canyons is like taking a trip into the dim, dark past, into the majestic virgin hardwood forest of the Southern Appalachians that was known to the American Indian and the earliest of our explorers. It is hard to believe that such a retreat, a haven with such feelings of isolation and wildness, can still be found within two hundred miles of major cities with a combined population of six millions.[1]

To reach the point at which such testimony could even be presented required a tremendous effort on the part of those involved in the original Sipsey campaign. Although that formal undertaking began in the late 1960s, its inspiration was born in 1892, in Coosa County, Alabama, in the person of Blanche Evans Dean.[2] An educator and self-taught field naturalist, Blanche Dean tirelessly cajoled all who would listen into an appreciation of Alabama's great natural heritage and gave them the means to do so. She founded both the Alabama Ornithological Society and the Alabama Wildflower Society, giving a statewide platform to outdoor enthusiasts. Working as a volunteer for the Birmingham Audubon Society, she led innumerable field trips, founded its Wildlife Film Series, and began its annual Outdoor Nature Camp for which academicians and field naturalists volunteered their time teaching about Alabama's ecology. Blanche was the author of several landmark botanical guides, including *Trees and Shrubs of the Southeast; Ferns of Alabama and Fern Allies;* and *Wildflowers of Alabama and Adjoining States,* coauthored with Amy Mason and Joab L. Thomas.[3] Her contemporaries credit her with nurturing an entire generation of knowledgeable Alabama environmentalists.

Blanche Dean was one of the first to advocate preservation of the Sipsey Fork of the Black Warrior River. In the 1940s she and Herbert McCullough, head of the Biology Department at Birmingham's Howard College, sought to have the area around Clear Creek Falls designated as a national park. Unfortunately, both the U.S. Army Corps of Engineers and Alabama Power Company coveted the area as a site for a reservoir, with the Power Company eventually winning out, flooding Clear Creek Falls under the waters of Lewis Smith Lake in the late 1950s. Before the flooding, Blanche Dean took as many people to Clear Creek Falls as she could, saying, "Remember this beauty. It will soon be gone."

Inundation of the Sipsey Fork helped focus the attention of Alabama's fledgling environmental community on the river's publicly owned remnant, its West Fork (commonly known as the West Fork Sipsey), flowing through the Bankhead National Forest north of Smith Lake. The West Fork's rich botany made it a natural attraction for Blanche Dean and for a special group of people who would devote themselves to Sipsey preservation, among them Mary Ivy Burks.

If one believes in fate, then surely Mary Burks was fated to become the mother of Alabama wilderness preservation. Passionate, tough, and resilient, a lover of all things wild and natural, Mary honed her communications and advocacy skills as a graduate in English from Birmingham Southern College, as a reporter for the *Birmingham Post,* predecessor to today's *Post-Herald,* and as a public relations specialist for a variety of volunteer organizations, including the Community Chest and the Birmingham Audubon Society. It was through Audubon that Blanche and Mary met. Blanche became Mary's "guru," introducing her to the splendors of the Bankhead National Forest. "I fell in love with the place," says Mary, and that was all the motivation that she needed.

In May 1967 Mary and her husband, Robert E. Burks Jr., invited Blanche Dean and several Birmingham Audubon Society activists to their home to discuss formation of a new, independent organization to focus exclusively on Alabama's environmental problems, which were (and still are) legion. This led ultimately to the creation of the Alabama Conservancy.[4] Although the new group immediately tackled several knotty environmental issues, notably air and water pollution, the fight to create the Sipsey Wilderness Area would define the Conservancy's early years.

In contrast to the vast public holdings of many western states, Alabama's public lands comprised only a small amount—about 5 percent—of the state's total land base. Its largest single block of public land lay in some 641,400 acres contained in national forest units scattered around the state: the Bankhead National Forest in northwest Alabama; the Talladega or Mountain Division of the Talladega National Forest in east Alabama; the Oakmulgee Division of the Talladega National Forest and the Tuskegee National Forest in central Alabama; and the Conecuh National Forest in the south.

Unlike national parks, with which they are often confused in the public mind, national forests are not purely recreational or ecological preserves but are managed by an act of Congress under a "multiple-use" concept, that is, "for outdoor recreation, range, timber, watershed, and wildlife and fish purposes."[5] Although management practices at the time gave at least cursory attention to each of these uses, the national forests were, in fact, under mounting pressure to be devoted primarily to commercial timber production.

Clear-cutting was the bugaboo that rallied the push for protection of public forests, in Alabama as well as elsewhere across the nation. Historically, national forests had been logged by means of "selective" cutting, which involved only the gradual removal of individual mature trees in any particular forest stand. Indeed, some areas, including parts of Alabama's national forests, were managed in an even more benevolent "custodial" fashion, which permitted forest stands to regenerate from abuses they had suffered in private hands before the creation of the National Forest System. In the Bankhead National Forest, the effect had been to restore the canyons of the West Fork Sipsey and its upper tributaries to a de facto wilderness condition.

In 1964, however, the U.S. Agriculture Department, which has jurisdiction over the national forests, issued a directive to its subagency, the U.S. Forest Service, to adopt clear-cutting, or the total removal of all trees in a stand selected for logging, as the standard practice in all the country's national forests. In the South, this involved removing large tracts of hardwood forest and replacing them with a monoculture of commercial pine. In 1969 congressional supporters of the logging industry introduced the National Timber Supply Act, later supplanted by the National Forest Timber Conservation and Management Act of 1969, which proposed to "increase substantially the timber yield from the commercial forest land of the Nation including that in the national forests . . . through intensified development and management."[6] The measure was defeated by strong national opposition, but it reflected the desire of the timber industry to "get at" the public lands.

By 1970 the national forests accounted for 40 percent of U.S. timber production, and a Nixon administration task force was recommending that national forest output be increased by 55 percent over the next eight years.[7] These changes, with the promise of more to come, and the resulting devastation in the national forests, sent shock waves through the nation's conservation community, but most especially in the eastern United States, where small remaining national forest wild areas were threatened with virtual annihilation.

Blanche Dean, Bob and Mary Burks, and the others who formed the Alabama Conservancy had no idea that they would soon be making history as their infant organization was thrust into the forefront of what would become the eastern wilderness movement.

CHAPTER TWO

It's Impossible

IN 1964 CONGRESS PASSED THE Wilderness Act "to secure for the American people of present and future generations the benefits of an enduring resource of wilderness. For this purpose, there is hereby established a National Wilderness Preservation System to be composed of federally owned areas designated by Congress as 'wilderness areas,' and these shall be administered for the use and enjoyment of the American people in such manner as will leave them unimpaired for future use and enjoyment as wilderness."[1]

Besides designating "instant" wilderness areas and specific wilderness study areas, the act instructed the Department of Agriculture, through its subagency the U.S. Forest Service, to conduct "reviews" of those lands within its jurisdiction that had been designated as "primitive areas," to determine "suitability or non-suitability for preservation as wilderness."[2] This directive effectively excluded what was then being called "de facto wilderness," famously defined by Sierra Club executive director David Brower as "[w]ilderness areas that have been set aside by God but which have not yet been created by the Forest Service.... [They are] the wilderness that sits in death row, ... and there has been nothing ... like ... a fair trial."[3]

The result was a nationwide citizen-based movement to identify national forest wild areas that deserved protection, regardless of the Forest Service's attitude toward them. Initially, these efforts centered in the western United States, but toward the end of the 1960s, under the forward-thinking leadership of the Wilderness Society, the movement began to spread east. Groups in two states, West Virginia and Alabama, were the first to respond, with Georgia soon to follow.

Charles S. Prigmore of the University of Alabama was among those attracted to the newly formed Alabama Conservancy. In its early days, the Conservancy was seeking a project or cause to propel it to a position of influence in the state, and Prigmore, a Wilderness Society member, felt that a forest preservation campaign would be just the ticket. He asked Mary Burks for a suggestion. "Charles," she said, "if there is any Wilderness [candidate] in Alabama, it's got to be in the Bankhead National Forest."

Alabama Conservancy mythology has it that, in 1969, Prigmore took two fellow University of Alabama professors, David T. "Tom" Rogers, an ecologist, and Joab Thomas, a botanist who later became president of the university, to the Bankhead Forest to identify an area to propose for preservation, and that the three produced a written report at the trip's conclusion. Interestingly, Tom Rogers does not recall the trip. Joab Thomas remembers the trip, but neither he nor Rogers recalls or has copies of the legendary report. Charles Prigmore is deceased, his records lost with him. However, a letter by Mary Burks to the Forest Service in December 1969 makes reference to a "report sent on September 4 [1969] by Dr. David T. Rogers and his associates in which they detail the unique natural assets" of the area, so perhaps the myth is true.

In any event, these three gentlemen were certainly the source of the very first Sipsey Wilderness proposal, which Tom Rogers describes as a "minimum" 6,000-acre area centered around the West Fork of the Sipsey River and the twin Bee Branch canyons, which had been administratively designated by the Forest Service in 1950 as a 1,200-acre "scenic area."

In the fall of 1969 Mary Burks, Tom Rogers, and others engaged the U.S. Forest Service in a series of meetings on the subject of a potential wilderness area in the Bankhead. Although organized into ranger districts with one or more offices in the local areas, all Alabama's national forests were under the direction of a single official, the forest supervisor, whose office was located in Montgomery. At the time, this position was filled by an individual named Del W. Thorsen.

One of the gatherings organized by the Conservancy involved a trip with Thorsen to the Bankhead itself. The group was composed of Alabama Conservancy activists and leaders of a variety of other organizations, including the League of Women Voters, the Audubon Society, and regional garden clubs. Ernest "Ernie" Dickerman, from the Wilderness Society's Washington, D.C., office also attended the meeting. Tom Rogers recalls that Forest Supervisor Thorsen was "extraordinarily antagonistic."

"He said, 'What do you want?' We said we wanted a minimum 6,000-acre Wilderness Area in the Bankhead National Forest. He said, 'No, that's not what you want. You want 6,000 acres today, you want 12,000 acres tomorrow, the next thing that you want is the whole Bankhead National Forest, and after that you want all of the National Forests in Alabama.'"[4]

Mary Burks says Thorsen was more than simply hostile. "He had no idea [why] anybody would want to do something like this. He didn't believe that we were real people. [He thought we were] some kind of exhalation out of the sewer. He said, 'It's impossible. There's no wilderness east of the Mississippi.' He said that there had to be a popular movement, and a formal study of the area, which they had no funds to perform, everything you could think of that was a reason why we couldn't do it. We said, 'We'll do all of that.'"[5]

The forest supervisor was taken aback by the group's preparedness, and in the end, he found himself agreeing to a joint study between the Forest Service and the Alabama Conservancy, and, remarkably, a moratorium on timber production in the study area while it was being prepared.

Thorsen asked for a formal written request. Mary responded by letter on December 9, 1969, and the forest supervisor found that he had been right in at least one respect—the proposal now was for an 11,000-acre area to be known as the Bankhead Wilderness. Tom Rogers says that since "we obviously had them on the run, we increased our proposal to 11,000 acres."

The study area was bounded on the south by Cranal Road, on the west by Kinlock Road, and on the north by Northwest and Bunyan Hill Roads. The eastern border was formed by a historic forest track called the Cullman Motorway, which ran north to south from Bunyan Hill Road to Cranal Road, and crossed the West Fork Sipsey at a long since washed-out ford just to the west of Fall Creek Falls. Bob Burks recalls that this boundary was not set farther east because of clear-cutting activity along the divides in that direction.

In a letter to Mary Burks, Thorsen agreed

to suspend certain Forest Service resource management activities for a period of one year on a reasonably sized area surrounding the Bee Branch Scenic Area. The area you have proposed (approximately 11,000 acres) is acceptable. Beginning December 9, 1969, the Forest Service will suspend the sale of timber (except salvage from insect, disease, or fire), reforestation, timber stand improvement and road construction on National Forest Lands within the 11,000 acre area for one year. . . . The Forest Service objective during this year will be to complete a Multiple Use Study on the area you have recommended for wilderness status.[6]

Thorsen could not bring himself to call the undertaking a "wilderness study," as his personal belief, as well as the position of his agency, was that nothing in the eastern United States qualified for protection under the Wilderness Act. But he deserves a star for prescience: during the course of the study, the Conservancy and the Forest Service agreed to expand the study area to some 15,000 acres of national forest land, extending the eastern border to Alabama Highway 33 to create "better geographic boundaries." The lower part of Borden Creek, which had been excluded by the first proposal, was now in the mix.

Conservancy participants began furnishing preliminary field reports as early as January 1970, but in true bureaucratic fashion, the Forest Service was much slower to respond. Frank Mayfield of the Regional Forester's Office in Atlanta indicated that the agency would not be able to act on the feasibility study until 1972, raising concern that the logging moratorium granted by Thorsen would be inadequate. The best that Mary Burks could get out of the regional office at the time was an assurance that "Forest Supervisor Thorsen intends to provide adequate time for a thorough and detailed study of the area."[7] The Sipsey later became the poster child for a new Forest Service "wild areas system," and the timbering moratorium was extended until Congress finally acted on the matter.

The leaders of the Wilderness Feasibility Study Committee organized by Mary Burks read like a who's who of Alabama's finest field ecologists, academicians and conservationists:

ornithology: Thomas A. Imhof of Birmingham, author of *Alabama Birds*

herpetology: Mike Hopiak and James Peavey of the Southeastern Herpetological Society

ichthyology: Mike Howell, Department of Biology, Samford University, Birmingham, and Don Dyckus, Alabama Department of Conservation

botany: Louise G. "Weesie" Smith and Blanche Dean, field biologists

geology: Denny N. Bearce, chairman, Department of Geology, Birmingham-Southern College

history: Dale Carruthers (Mrs. Thomas N., Jr.), amateur historian and volunteer researcher

speleology: James and Fran Alexander of the Huntsville Grotto of the National Speleological Society

game wildlife: Charles Kelley, director of the Game and Fish Division of the Alabama Department of Conservation, and Ralph Allen, also of the department nongame wildlife: Dan Holliman, Department of Biology, Birmingham-Southern College

trails and chairman of the field parties: James Manasco, outdoors enthusiast

Against such an array, the Forest Service didn't stand a chance. The reports these experts and their volunteer assistants produced established beyond question the ecological significance of the study area. And a couple of results produced by the joint study influenced subsequent legislation, both for the initial establishment of the Sipsey Wilderness and for its proposed enlargement in the 1980s.

The first came from the Forest Service itself, in a soil management report compiled as a part of the joint undertaking. This study concluded that 65 percent of the study area would suffer "moderate to severe" impacts in terms of "soil erodibility" and "potential for watershed damage with intensive forestry management." Fifty-six percent of the study area fell into the "severe" category.[8]

The importance of this finding was underscored in the field report on geology prepared by Denny N. Bearce at Birmingham Southern College:

> The [Sipsey] canyons are not self-sufficient, however. They obtain their water supply from slope wash and subsurface seepage from the divides around them. If these divides are stripped of trees pursuant to the even-aged forestry management policy, the canyons will suffer. Loss of plant life on the divides will lead to erosion. . . . The delicate complex faunal and floral systems within the canyons could well be washed out in a succession of . . . floodings. In dry periods the inability of the barren divides to hold water and thus to yield a steady supply of subsurface drainage to the canyons below will place the plant and animal life therein at the mercy of the weather. Small, shallow-rooted plants and moisture-needing animals will not survive.
>
> Geology has provided a natural refuge system in the Bankhead Forest. This system found in the deep canyons can serve and help to sustain life in the surrounding forest, but in turn the surrounding forest is needed to protect and sustain it.[9]

These conclusions were frequently cited to rebuff the efforts of opponents both in the Forest Service and in Congress to restrict any protected area to the canyons alone. At issue was preservation of the unique watershed of the West Fork Sipsey River.

With the field studies under way, Mary Burks and the others turned their attention to the politics of wilderness preservation. Today, Mary says, "We didn't have any idea what we were doing. We learned, believe me. We learned the hard way." While this was certainly true to a point— these folks had never conducted a grass-roots campaign before—they had good instincts. One of the things they did was launch a petition drive that first, called people's attention to the issue, and second, procured some twenty-three thousand signatures to be presented to public officials at the appropriate time.

Elberta Gibbs Reid, an early Conservancy board member and long-time activist with the Birmingham Audubon Society, credits Mary Burks with the success of the public campaign. Mary served as the first president of the Alabama Conservancy, then became its wilderness chairman, and in 1971, its first executive director, leading the Sipsey Wilderness campaign throughout. But she wasn't the only effective leader. Louise "Weesie" Smith, who became the third president of the Alabama Conservancy, "not only knew a lot of people" with influence, says Elberta—"but also was well known and respected in her own right—a good one to have as a supporter." Weesie Smith was recognized as one of the nation's finest amateur field botanists, in the mold of Blanche Dean.[10]

In northwest Alabama, Jim and Ruth Manasco were the most effective local advocates, as was one Rickey Butch Walker. Bob and Mary Burks tell of early Bankhead trips in which they encountered numerous hand-made signs posted around logged areas. She absconded with a few, which she showed me.

One features an angry and cursing Smokey Bear, declaring, "I QUIT (*@&*$*%@* *Censored*) The Forest Service is Doing More Damage Than Forest fires."

Another, titled "Where Have All the Oak Trees Gone?" depicts a starving turkey and rabbit saying, respectively, "I could never eat pine cones" and "These pines just kill me."

A third shows deer, turkey, and squirrel gathered around a sign in front of a clear-cut that reads, "This Area is Clearcut to be Planted in Pine Trees. All Wildlife is banned by order of the U.S. Forest Service." Then, "WHAT DO YOU WANT—Wildlife or Pine Trees?"

Mary says, "We were struck with admiration. The Forest Service said they were going to get the FBI to track down the people who were doing all this." When Mary heard that Butch Walker was the culprit, she was afraid to mention it or discuss it, for fear she would be called to testify against him. The

Forest Service never succeeded in their hunt for the sign maker, if they conducted such a search at all, and Butch Walker today remains a vocal Bankhead advocate and leader in the group known as Wild Alabama.

And then there was the Birmingham Audubon Society icon Walter F. Coxe. Born in Hall River, N.C., in 1898, and a graduate of Georgia Tech, Walter moved to Birmingham in 1940 to pursue a career in the advertising business. An indefatigable volunteer, he served as president of both the Birmingham Audubon Society and the Alabama Ornithological Society, and was active with Civitan and the Boy Scouts. More to the point, he was an ardent advocate of the wild, a spellbinding, much sought-after speaker on nature. Walter Coxe became a well-known local personality, says Elberta Reid, as a regular on the early-morning Birmingham television program *The Country Boy Eddie Show:* "This was before cable [TV] and nobody had nature films on television, so Walter [would appear on *Country Boy Eddie*] and say, 'Come to the [Audubon Society's] wildlife films,' and everybody would come. Everybody seemed to know him. People would ask for his autograph in country stores—like a movie star. Walter was a marvelous person to spread the [Bankhead] message. He really got people on his side."

To spread the message, Walter teamed up with photographers Perry Covington and Dennis Holt and editor Elberta Reid to produce a film, *The Bankhead Forest—An Alabama Adventure,* which Walter narrated and then took on tour around the South as part of the National Audubon Society's Wildlife Film Series. For this, and for his decades of contribution to environmental advocacy, Walter Coxe was given a 1984 Sol Feinstone Environmental Award by the State University of New York, the first of three Alabama wilderness activists to win the honor. (The other two were Mike Leonard and me, for efforts chronicled in forthcoming chapters.)

The Sipsey campaign had one other star in its firmament, and he called himself a pro. "There were a lot of people without whom there would never have been this Wilderness," says Elberta Reid's husband, Robert R. Reid Jr., "certainly including Mary Burks. But it is pretty clear that it wouldn't have been there without Charles Prigmore."

Prigmore, who succeeded Mary Burks as president of the Conservancy, taught courses in public policy and political science at the University of Alabama, but he was also a professional lobbyist and had represented various interests before Congress and the Alabama legislature. Bob Burks says that Prigmore was known for trying "to find something on the [politician he was lobbying] and let him know that he knew, and get his attention. He would go to the same 'watering holes' as the state legislators, to see who they came in with, who their girlfriends were, and then mention it to them the next day." Elberta Reid says simply, "He was an act to watch."

Lobbying the Sipsey legislation through to fruition was to be Charles Prigmore's task. "Charles was extremely knowledgeable," says Elberta Reid. "He always knew what to do, where to go, who to see. He was full of energy and decision. Charles taught us all how to lobby."

The Sipsey team in place—and a remarkably competent and professional one it was—the leaders now turned to confronting the U.S. Congress and the monolithic, hostile bureaucracy of the Forest Service.

CHAPTER THREE

Purity

To BECOME A UNIT OF THE National Wilderness Preservation System, an area must be designated by an act of Congress. This means that at some point politicians must be involved, for better or for worse. In the case of the Sipsey, four Alabama public officials held the key to the first gate on the path to preservation.

The Bankhead Wilderness Study Area straddled the Winston-Lawrence County line, which placed it in two congressional districts.[1] Winston County was represented by Tom Bevill, Democrat of Jasper, but the majority of the land lay in Lawrence County, the domain of Congressman Robert E. "Bob" Jones, Democrat of Scottsboro. The other two key officials were Alabama's U.S. senators, John J. Sparkman and James B. "Jim" Allen, both Democrats.

Congressional courtesy demands that anything affecting a particular district must begin with, or at least have the tacit concurrence of, the congressman representing that district, so the first contacts Mary Burks made were with Bevill and Jones. Bevill wrote, "Thank you very much for your report about developments of the establishment of a wilderness area in the Bankhead National Forest, which I received today. I want to assure you that I will support this plan to establish the Bankhead Wilderness. If there is anything you would like me to do, please do not hesitate to call upon me."[2]

Bob Jones, however, was noncommittal, a portent of things to come: "I am greatly interested in this proposal and would appreciate being kept advised of future progress relating to this proposition. Although I do not know the total requirements of what such a proposal would involve, I certainly will make an effort to look into this matter."[3]

One other Alabama congressman began taking a particular interest in the Sipsey, and that was John H. Buchanan Jr., Republican of Birmingham. His involvement seemed natural, since so much of the support for the wilderness was being generated by his constituents. Buchanan took the time to meet personally with Forest Supervisor Thorsen during the Christmas recess of 1969 to apprise him of his interest.

Thorsen, meanwhile, was on his way out. In February 1970 the regional forester's office in Atlanta advised the Conservancy that Thorsen was being "promoted and transferred" to the Francis Marion–Sumpter National Forest in South Carolina and would shortly be replaced by one John Orr, a graduate of the University of Georgia. Regardless of who the forest supervisor was going to be, however, Sipsey supporters would have to surmount the hostility of the Forest Service and their allies in Congress toward the concept of wilderness in the eastern national forests. Regional Forester T. A. Schlapfer stated it this way:

> Each area proposed for inclusion must qualify under the criteria established by the [Wilderness] Act which states in part . . . "Where the earth and its community of life are untrampled [*sic*—the word is actually "untrammeled"] by man, where man himself is a visitor who does not remain."
>
> The Chief of the Forest Service, the Secretary of Agriculture, as well as the responsible congressional committees have steadfastly opposed inclusion of any area into the National System unless it absolutely meets the established criteria.[4]

In September 1970 the Conservancy sent a group to Washington to meet with the members of the Alabama congressional delegation, congressional committee members and staffers, and the national office of the Forest Service to push their Bankhead wilderness proposal. "We mainly got cold shoulders from everyone," recalls Elberta Reid. "If there is a footprint of man, then it won't qualify."

Mary Burks recalls meeting with one of the high officials in the recreation division of the Forest Service. "We invited him to come see the Sipsey. He drew himself back and said, 'I don't have to go to Alabama to know that there is nothing in Alabama that would qualify!' And so, we thought, we'll show him."

And show him they did, but it would require years of effort and a national movement to prevail over this attitude, which came to be known as the "purity" doctrine. It was asserted at the outset not just by the Forest Service but also by certain national conservation groups that took the position that admitting second-growth forests in the East into the wilderness system would somehow "degrade" it.

The purity doctrine had been in practice within the Forest Service, in fact if not by name, for a number of years, but not until September 1971 was it publicly enunciated at the highest levels of the agency. Speaking before the Sierra Club's Biennial Wilderness Conference in Washington, D.C., Associate Chief John McGuire proclaimed, "The areas with wilderness characteristics as defined in the Wilderness Act are virtually all in the West."[5]

Douglas W. "Doug" Scott, who was with the Wilderness Society at the time, says that McGuire's pronouncement was a "deliberate misinterpretation of the 1964 Wilderness Act."[6] Scott says,

> By the early 1970s the agency was feeling seriously threatened by growing grassroots pressures for Congress to protect de facto wilderness on eastern national forests. Agency leaders talked themselves into the conclusion that no areas in the East could even qualify as wilderness under the Wilderness Act. In no small part were they motivated by the desire to sustain their unwarranted *purity* interpretation . . . in the East, in order to apply it as a means of minimizing the expanse of wilderness boundaries in *western* national forests, where lower elevation areas also often had a history of some past development.[7]

Unfortunately, the Forest Service's bias was being seconded by large and influential outdoor clubs in the Northeast and by the Sierra Club and the Izaak Walton League, all of which were primarily concerned with overuse of designated wilderness areas. They argued that admitting eastern areas would "lower the standards" of the wilderness system, resulting somehow in encouraging "the growing and irresistible pressure of recreationists themselves who, with snowmobiles, outboard, ATV and other gadgetry, or sheer numbers lean their weight against every wilderness boundary."[8]

Ernie Dickerman, Doug Scott, Rupert Cutler, and other Wilderness Society staff members worked diligently during the early 1970s to overcome this mind-set and bring these national organizations into the eastern wilderness fold. "In our opinion," said Dickerman, "it is preferable to put into the System an area which may contain some minor work of man than it is to reject the entire area or a significant portion of it in order to avoid such minor features. After all, the objective is to preserve wilderness, not seek reasons for rejecting its preservation. Once an area is in the system, however, we can expect to fight with maximum skill and diligence to prevent even the most minor sort of intrusion into a legally designated wilderness area."[9]

The debate over the legitimacy of the purity doctrine would extend even to the ranks of Forest Service personnel. In his history of the National Forest

Wilderness movement, former U.S. Forest Service national historian Dennis M. Roth quotes a presentation made by an agency ecologist: "Nature is a great healer and in just a few years imprints of man's work will become substantially unnoticeable. . . . The argument is made that Wilderness, by definition, can never be restored. In actual fact, there are many examples where, for all practical purposes, Wilderness has been restored. It is just a matter of how much time is involved."[10]

Others of a legalistic bent maintained that the agency's reliance upon the defining words in the act, "where the earth and its community of life are untrammeled by man," ignored the "further definition" of wilderness, contained in the same section: "An area of wilderness is further defined to mean in this Act an area of undeveloped Federal land retaining its primeval character and influence . . . [which] *generally* appears to have been affected primarily by the forces of nature, with the imprint of man's work *substantially unnoticeable.*"[11]

But, in the end, it would not be for the Forest Service, or the Sierra Club or the Wilderness Society, to settle the purity dispute, but the Congress of the United States, since it created the wilderness system in the first place. There, proponents of eastern wilderness would find support in the legislative history of the Wilderness Act, and more importantly, from key members of Congress. Perhaps none was more influential than Sen. Frank Church of Idaho, who was floor manager for the act when it was passed by the Senate. He pointed out, "[The 1964 Wilderness Act] placed three eastern areas into the National Wilderness Preservation System [that] . . . had a former history of some past land abuse. This was . . . a standing and intentional precedent to encourage such areas to be found and designated under the act in other eastern locations."[12]

The three areas mentioned by Senator Church were the Shining Rock and Linville Gorge Wilderness Areas in North Carolina and the Great Gulf Wilderness in New Hampshire. Shining Rock offered a convincing example of what the senator was talking about, as it had a "history of extensive railroad logging—and a huge logging slash fire—between 1906 and 1926, when it became national forest land." There was even more recent logging under Forest Service jurisdiction and a timber sale contract outstanding at the time Shining Rock was being considered for inclusion in the National System. The House Interior Committee reviewed each of the areas proposed for "instant" wilderness in the 1964 legislation, including Shining Rock, to insure that they met the criteria for the new act.

Another example of congressional intent was the Great Swamp Wilderness in New Jersey ("just 30 miles from Times Square"). The third wilderness area designated by Congress after passage of the Wilderness Act, it contained "a two-lane paved road with ditches, shoulders, several bridges, and several suburban homes on private inholdings." The road was closed after the area was designated as wilderness.

In sum, said Senator Church, "Nothing could be more contrary to the meaning and intent of the Wilderness Act [than the purity doctrine]. The effect of such an interpretation would be to disqualify almost everything, for few if any lands on this continent—or any other—have escaped man's imprint to some degree."

These were the arguments that would shape the eastern wilderness debate, but in the fall of 1970, when Alabama Conservancy leaders were in Washington to lobby their proposal, the discussion had barely begun, and purity was heavily entrenched in the policies of the federal government. It is all the more remarkable, then, that our Alabamians received such a warm reception for their wilderness from Sens. John Sparkman and Jim Allen.

Asked why Sparkman was willing to sponsor an eastern wilderness proposal, knowing that it would attract opposition from the Forest Service and its congressional supporters, including key committee chairmen, Mary Burks says that the senator told her he had never seen a popular movement "catch fire" like the environmental movement of the time. As for Senator Allen's interest, Elberta Reid credits one of his young staff assistants, Wendell Mitchell of Luverne, who would later become a powerful member of the Alabama legislature. "He was the one who got Jim Allen's attention, and got us in the door," she says. "He made him see the wisdom of doing this."

When the first session of the 92nd Congress convened in early 1971, Senator Sparkman advised Charles Prigmore, "I shall be happy to introduce legislation to establish the Bee Branch Wilderness area," adding that he would seek assistance from Ernie Dickerman and the Wilderness Society in drafting the legislation.[13] Meanwhile, aided by the commissioner of Alabama's Department of Conservation and Natural Resources, Claude D. Kelley, Mary Burks obtained Gov. George Wallace's endorsement of the proposal. Wallace wrote to Alabama forest supervisor John Orr: "It appears to me that the best interest of the public will be served if the proposed wilderness area, comprising approximately 10,000 acres, could be sanctioned by the U.S. Forest Service.... I am convinced that the establishment of this area in perpetuity would be a wonderful gesture on the part of the U.S. Forest Service and certainly would give additional credibility to your agency's multiple use concept which is now in existence."[14]

Charles Prigmore reported to Mary,

> Things are happening fast on the Bankhead.... Sen. Sparkman ... will introduce. This is really better than having Sen. Allen take the leadership, since Sen. Sparkman is a Chairman of one of the top committees. Nobody will oppose him on this issue, and I even suspect that [Congressman] Bob Jones will come around. They are very close, I believe.
>
> Gov. Wallace will announce his support soon, and you should wait until then to spread the word widely about Sen. Sparkman (since Gov. Wallace likes to be first).[15]

Governor Wallace did "go first," and then, on April 21, 1971, Sen. John Sparkman introduced in the U.S. Senate bill number S. 1608, proposing a "Sipsey Wilderness Area ... comprising about twelve thousand acres." Jim Allen cosponsored the bill, as did nine other senators from around the country. Supporters hoped that the measure would be referred to the Senate Committee on Interior and Insular Affairs, which was chaired by one of the Sipsey cosponsors, Sen. Henry M. Jackson of Washington, and on which three other cosponsors—Lee Metcalf of Montana, Mark Hatfield of Oregon, and Frank Church of Idaho—served. But Forest Service allies saw to it that the Senate Agriculture Committee exercised its traditional jurisdiction over eastern national forests and had it assigned there, where its success was extremely doubtful.

No "official" map survives that depicts how the wilderness proposal became "about twelve thousand acres," but Alabama Conservancy records indicate that the boundary of its original 11,000-acre proposal was pushed eastward so as to encompass the lower part of Borden Creek. From the 15,000-acre Bankhead Wilderness Study Area, there were now substantial exclusions to the east and northwest of Borden Creek so to as to eliminate recent clear-cuts and a few private inholdings, as well as a large triangle in the southwestern part of the study area to exclude inholdings. The end result approximated 12,000 acres.

Congressman Bob Jones "was infuriated," says Mary Burks, that Sparkman had proceeded without his consent or involvement, and Congressman Tom Bevill was unhappy as well. Charles Prigmore's strategy of asking Senator Sparkman to take the lead appeared to be backfiring.

In May 1971 Jones and Bevill struck back with House legislation of their own. H.R. 8739 sought to create a Sipsey National Recreation Area of some 11,330 acres, surrounding a Bee Branch Wilderness of 1,240 acres that was composed, roughly, of the existing Bee Branch Scenic Area. The bill authorized the Forest Service to manage the recreation area "in accordance with the laws, rules, and regulations applicable to national forests," which meant that the agency had wide discretion over what activities to permit and was under no meaningful statutory mandate to preserve the land in its wild state.[16]

Moreover, the tiny size of the proposed Bee Branch Wilderness was far below everyone's accepted acreage standard for national forest wilderness—5,000 acres minimum—making it likely that the bill would attract national opposition for this reason alone.

This development flew in the face of the remarkable outpouring of support that was being generated back in Alabama for protection of the Sipsey as a statutory wilderness. State representative Ben Erdreich of Birmingham, who would later represent the city in Congress, had sponsored and secured passage of a joint resolution of the Alabama legislature in favor of the Sipsey Wilderness. When Jones and Bevill introduced their alternate legislation, Erdreich wrote to the *Birmingham News:* "I am appalled by the introduction of legislation in Congress to create this [recreation area] which, apparently, sets aside only a small portion as true wilderness. . . . [I am] not at all pleased with the apparent subterfuge of the legislation involved, and [hope] that the Bill submitted will be amended so as to create a true Wilderness Area comprising the entire approximately 11,000 acres."[17]

"The Jones-Bevill bill is worse than no bill at all," said the Alabama Conservancy's press release, noting that Governor Wallace, the state legislature, some one hundred organizations around the state and several of the major newspapers had all formally endorsed the Conservancy's proposal. "Everyone wants a full 11,000-acre wilderness."[18]

The *Birmingham News* weighed in with a passionate editorial:

"As far as protecting . . . one of the few remaining real wilderness areas in the nation is concerned, the Jones-Bevill bill's provisions would be about as effective as a box of matches in the hands of a pyromaniac or turning a 'cut and move out' lumbering outfit loose in the area for a few weeks. . . . Pinching in the [designated] wilderness area to no more than one-tenth of its originally considered size, and surrounding it with roads, camp sites and other appurtenances far removed from the idea of the undisturbed quietude of virgin forests could not but help result in destroying the idea of preserving our remaining natural resources as a wilderness. . . . The Jones-Bevill bill would negate the efforts of the [Alabama] Conservancy and the wishes of thousands of other Alabamians. It would not further the efforts of conservation. It would help defeat them."[19]

The uproar attracted the attention of the U.S. postmaster general, Winton Blount of Montgomery. Blount was entertaining the notion of running against John Sparkman for the Senate, and he apparently saw the Sipsey campaign as a potential vote-getter. The *Birmingham News* reported:

Accompanied by Forest Service officers, newsmen and members of the [Alabama] Conservancy, Blount plunged through great dark forests of oak and hemlock and holly, sliding and slipping down the deep canyon sides and past great cliffs to the bottom, where the unspoiled Sipsey winds along in a perpetual green gloom.

In one area, so hushed and removed that the crisp rustle of falling leaves could be heard on the autumn air, Blount raised his head and said, "You hear that?" No one heard anything, and that was his point. "You don't hear that in the city," he said. "That's quiet."

... Blount's suggestion was to introduce a new law which would circumvent the U.S. Forest Service's wilderness interpretation but, in effect, give the Sipsey area full "wilderness" protection.[20]

Congressman Jones grew defensive. A letter he began circulating in response to the protest reads, in part, "Unfortunately, the opportunity for preserving a large amount of wilderness area in Alabama passed before either you or I came on the scene.... [The Sipsey] lacks that critical factor of being virgin forest, retaining its primeval character, untrammeled by man.... You may be interested in reading a letter from the Forest Service concerning this matter, which is enclosed."[21]

The "letter from the Forest Service" referred to by Jones was from E. W. Schultz to Congressman Tom Bevill: "Most of these 15,000 acres [in the Bankhead study area] have been acquired from private ownership within the past 50 years.... Except for about 400 acres in Bee Branch, most of the area has been logged and farmed. Many of the old logging road grades are still visible.... By comparing [this] information with suitability criteria of the Wilderness Act, it appears that the [Sipsey] does not retain its primeval character and influence and does not meet the definition of wilderness."[22]

Not surprisingly Sipsey supporters viewed the Jones-Bevill bill with alarm. They had experienced a very hostile reception from Bob Jones in particular. Mary Burks says that if Jones "heard we were in the halls of Congress, he would go out the back way, get on the elevator, and leave so that he wouldn't have to deal with us." Jones's motivation seemed highly suspect at the time.

Viewed with thirty years' hindsight, the Jones-Bevill bill might seem today more understandable, given the opposition of the U.S. Forest Service and its allies in the Nixon administration to eastern wilderness. In July 1971 Under-Secretary of Agriculture J. Phil Campbell appeared at a Republican Party fundraiser in Montgomery and proclaimed that the chances of getting a designated wilderness in the Bankhead were "zero": "In fact, said Campbell, there would be virtually no areas in the Southeast that could qualify as a 'wilderness area' under the definitions set up by Congress. To be eligible, he noted, the original trees and wildlife 'can't have been tampered with but must have been virtually untouched through the years.' Almost all timberlands in the Southeast, he explained, have 'been cut over several times since the white man took them over. Bankhead,' said Campbell, 'doesn't come close to being in its original state.' "[23]

Jones and Bevill saw their legislation as being less likely to attract administration opposition, and were motivated, at least in part, by this. Jones told the press: "Our objective . . . and the objective of the Alabama senators and the Alabama Conservancy is basically the same—to set aside and preserve the Bee Branch area of the forest. Our House bill is not set in concrete. We are willing to consider amendments that might improve our bill or win the wide support that any such bill must have to get approval of both House and Senate and survive a possible veto by the President."[24]

So, while this approach might seem reasonable, given the context of the times, it was, in the final analysis, one that merely appeared to protect the land rather than actually doing so. In that respect, it was at worst a cynical politician's trick, or at best, a ratification of Forest Service intransigence. Congressman Jones, in particular, could not refrain from revealing his own sympathy for the agency's position: "Congress has the authority to ignore Forest Service criteria and declare any area a wilderness area, but the mere fact of declaration would not make it so."[25]

For his part, Charles Prigmore was growing so mad at Forest Service opposition that he was moved to write Pres. Richard Nixon with a somewhat silly threat. Proclaiming the Forest Service to be "arrogant and indifferent to public concern," Prigmore told the president, "If you can't stop the U.S. Forest Service, then we will turn to our own leader in Alabama, George Wallace, for help. He has shown his willingness to stand up against federal bureaucracies and take the part of the common man against the powerful coalition of bureaucrats and ruthless exploiters of our natural resources."[26]

"I really like politicians," Prigmore told Bob Burks, "but I hate bureaucrats."

Emotions aside, however, Prigmore, Mary Burks, and the other Sipsey leaders knew that perpetuating the disagreement would only result in the failure of any legislation at all, so a peace conference was arranged. In September 1971 the Alabama Conservancy sent a delegation of its board members to meet with Sparkman, Allen, Jones, and Bevill. The Conservancy contingent included Mary Burks, Charles Prigmore, Elberta Reid, Mike Hopiak of Birmingham, John B. Scott Jr. of Montgomery, and Rod Jenkins of Huntsville. The result was an agreement in principle, with precise acreages still to be worked out, for most of the 12,000 acres in the Sparkman-Allen bill to be designated wilderness, while the smaller remainder, situated around the perimeter of the new wilderness, would receive the "recreation area" status contemplated by the Jones-Bevill legislation. Sen. Jim Allen, as a member of the Senate Agriculture Committee, would take the lead in getting the new measure approved by the upper chamber, and Jones and Bevill would get House approval for whatever the Senate passed. All that remained was to allocate the acreages and attempt to get the Forest Service to accept the idea.

Returning to Alabama, John Scott undertook to devise a proposal to split the land between wilderness and recreation area. Scott was chosen for the task primarily because of his position as a respected Montgomery attorney. He observed that "I know less about the Bankhead than anyone concerned, and personally have no strong feelings about the divisions" to be made.[27]

Nevertheless, working with a Forest Service employee, one Fritz Behrins, John Scott came forth with a proposal for roughly 9,000 acres as wilderness and 3,000 as recreation area, the latter divided into three "enclaves." The largest part of the recreation area encompassed lower Borden Creek and the land around Fall Creek Falls. "The other two enclaves were [on the southern and western perimeter of the wilderness and were] just arbitrarily drawn to provide some substantial recreation areas in portions of the tract as far removed from the main gorges as possible."[28]

Subsequent protests from within the wilderness ranks resulted in Borden Creek and Fall Creek Falls being restored to the wilderness allocation and a smaller recreation area enclave to the east of Bee Branch was substituted. The adjusted acreage totaled approximately 11,000 acres. All this work and worry, however, soon took a backseat to a new national offensive by the U.S. Forest Service to deal with these pesky eastern wilderness proposals.

CHAPTER FOUR

A Seething Mass of Happenings

WHILE ALABAMA CONSERVATIONISTS AND their elected officials struggled for a solution to the conflicting Bankhead National Forest proposals, the U.S. Forest Service decided the time had come for it to regain the initiative. As Dennis Roth explains,

> In 1970, members of the West Virginia congressional delegation introduced several wilderness bills, as did members of the Alabama delegation for the Sipsey area on the Bankhead National Forest. These bills did not get out of committee, but they showed the Forest Service that public interest in eastern wilderness was mounting. . . . The Forest Service realized that the public would not be satisfied with administrative protection [for the wilderness] and that if the Forest Service did nothing, public "impatience may ultimately preempt Forest Service leadership in this area."[1]

In the summer of 1971 the regional foresters headquartered in Milwaukee and Atlanta came forth with a proposed statutory alternative for eastern national forests, to be known as the Wildwood Heritage System, a title that was later dropped in favor of "Wild Areas." The agency decided the Sipsey would be the prototype for the new designation and asked the Alabama Conservancy to return to Washington in November 1971 for a meeting sponsored by the Izaak Walton League, which was supporting the idea. It was a reluctant Alabama delegation that attended. "Once again [the Forest Service] demonstrated contempt for our concern and indifference to the time and money such meetings cost the citizen," Mary Burks complained to Senator Sparkman.[2]

In addition to Forest Service personnel from the Washington, Atlanta, and Milwaukee offices, the conference was attended by the executive directors from the national offices of the Izaak Walton League, the Wilderness Society, and the Sierra Club, as well as representatives of Friends of the Earth and the American Forestry Association. Charles Prigmore, Mary Burks, and Weesie

Smith represented the Conservancy. The Forest Service presented draft legislation for critique by the group.

The Forest Service attempted to distinguish the Sipsey from the agency's concept of wilderness. The draft's preamble recognized the "near-natural character of the Sipsey area" and "its ability to provide . . . needed primitive forms of recreation . . . in a near natural setting." It required that the wild area be managed so that its "primitive, near-natural conditions shall be maintained, restored, and protected. . . . Maximum public use and enjoyment shall be encouraged. . . . Commercial and other uses shall be permitted where necessary and in accord with the purposes of the Act. . . . Timber cutting may be practiced to restore or maintain near-natural conditions . . . but not for commercial purposes."[3]

Mary Burks wearily dismissed the whole thing as just another "recreation area—something the Jones-Bevill bill has already done."[4] However, she and the other Conservancy leaders knew they could not permit the Forest Service to assume, or to represent to congressional members, that the Conservancy acquiesced in this Wild Areas proposal; so, returning to Alabama, she and Weesie Smith completely rewrote the proposed legislation, modeling it after the Wilderness Act, but omitting its definition of "wilderness," so that the Forest Service could not again raise its purity objections. The draft was distributed to the participants of the November meeting and to the affected members of the Alabama congressional delegation.

In February 1972 President Nixon released his annual environmental message to Congress, which seemed to endorse the concept of both eastern wilderness and wild areas:

> One of the first environmental goals I set when I took office was to stimulate the program to identify and recommend to the Congress new wilderness areas. . . . A few of these areas proposed today or previously are in the eastern section of our country, but the great majority of the wilderness areas are found in the west. This of course is where most of our pristine wild lands are. But a greater effort can still be made to see that wilderness recreation values are preserved to the maximum extent possible in the regions where most people live. . . . I am therefore directing the Secretaries of Agriculture and the Interior to accelerate the identification of areas in the eastern United States having wilderness potential.[5]

The Forest Service chose to downplay the reference to "wilderness potential" in the message, and to focus instead on the phrase advocating preservation of "wilderness recreation values," which the agency interpreted as encouraging a wild areas system rather than statutory wilderness. J. Wayne Cloward, assistant regional forester in Atlanta, said, "As a result . . . and apparent strong demand from many sources, we [the Forest Service] began to put together a concept for a system that would provide similar opportunities [in the East] for primitive type recreation that are now provided for in the West by the 1964 Wilderness Act. . . . The Wild Area concept . . . began as a result of our trying to define an alternative to statutory wilderness for the Sipsey River."[6]

Sen. John Sparkman decided to take the matter in hand. On February 23, 1972, without involving or consulting Sipsey supporters at home, Sparkman and Jim Allen withdrew their previous wilderness legislation and introduced two new bills. S. 3224 sought to establish a Sipsey Wilderness of 6,000 acres and a Sipsey National Recreation Area of 3,000 acres—a proposal that essentially tracked the "agreement in principle" reached back in September 1971, but with a total acreage reduction down to 9,000 acres, something that completely mystified the Alabama Conservancy.

The same day, Sparkman and Allen also introduced S. 3225, the Sipsey Wild Area Act, establishing a Southeastern Wild Areas Preservation System of federal lands "to be administered . . . so as to provide for the protection of these areas [and] . . . the preservation of their wild character." The legislation mandated a review of all southeastern national forests for wild areas that might qualify for inclusion in the new system, and designated a Sipsey Wild Area of 9,000 acres for immediate enactment. Ironically, the language of S. 3225 was virtually identical to the wild areas counterproposal that Mary Burks and Weesie Smith had submitted to the Forest Service the previous year. Both bills were referred to the Senate Agriculture Committee, where Jim Allen was a member.[7]

Then, on June 13, 1972, Sen. George Aiken of Vermont and Sen. Herman Talmadge of Georgia introduced their own wild areas bill, S. 3699, the National Forest Wild Areas Act of 1972. This called for the creation of a wild areas system for federal lands "in the Eastern United States," but it did not propose "instant" designation for the Sipsey or any other particular area. Its provisions resembled those in the Sparkman-Allen Wild Area bill, but it had more detailed instructions for a review of federal lands, requiring public hearings and input from the states, and other procedural details. S. 3699 was actually Senate companion legislation for a bill introduced in April in the House of Representatives, H.R. 14392, by Congressman John Kyle of Iowa. Then, on June 30, 1972, Congressman Kenneth Hechler of West Virginia introduced two additional versions of a National Forest Wild Areas Act, H.R. 15651 and H.R. 15851, thoroughly confusing matters.

The result was bedlam, nationally and locally.

Wild areas deepened the rift between the Wilderness Society, which wanted eastern wilderness instead, and such groups as the Izaak Walton League, which still clung to the purity doctrine, and the Sierra Club, whose leadership felt that wild area designation might provide the best opportunity for protection in light of the Forest Service's unyielding hostility to eastern wilderness. Helen McGinnis was a Sierra Club activist from West Virginia, as well as a member of the West Virginia Highlands Conservancy, a grass-roots organization that, like the Alabama Conservancy, was among the first to push for wilderness designation in their national forests. McGinnis believed that many Sierra Clubbers felt that "the Forest Service is too powerful to oppose . . . and if they prefer wild area to wilderness, well they're the experts." These folks, too, saw wild areas as meeting the "need for legislation to protect other areas not qualified as wilderness."[8]

Compounding the confusion was a struggle going on in the U.S. Senate about jurisdiction over wilderness, and now, wild area, proposals. Dennis Roth says,

> The Senate Interior Committee had handled all previous wilderness bills dealing with western forests and thus claimed jurisdiction over eastern wilderness bills. The Agriculture Committee had jurisdiction over the national forests in the East because their lands had been purchased from private owners rather than reserved from the public domain like those in the West. Therefore, its members also asserted jurisdiction over eastern wilderness bills. The institutional rivalry was paralleled by a personal coolness between the chairmen, Senator Herman Talmadge of Georgia [Agriculture] and [Sen. Henry M. Jackson of Washington, Interior], who communicated only through their aides.
>
> The Wilderness Society was distressed as much by the possibility that the Agriculture Committee might control eastern wilderness bills as it was by the possibility of a separate wild areas system. . . . [Agriculture] Chairman Talmadge professed to have more faith "in the management plans being drawn up by the Forest Service to provide true multiple use . . . than on designations forced upon the Forest Service by strident interest groups."[9]

In Alabama, Charles Prigmore was suddenly struck with the fire of compromise. Only a few months after characterizing the Forest Service to the press as "arrogant and indifferent to public concern," and threatening to have

George Wallace "sic 'em," Prigmore was moved to lecture a meeting of the Sierra Club in Tuscaloosa:

> The concern of many conservationists is that we do a great disservice to our own cause by stereotyping all Forest Service people as untrustworthy, responsive only to timber interests, rigid in their adherence to bureaucratic constraints and opposed irrevocably to the conservationists' position. The cause of wilderness preservation therefore runs the risk of a serious backlash, or at the very least, a loss of national confidence, if it does not evidence a willingness to compromise, to balance the various national interests, to seek an equilibrium in the clash of values, to act, in short, in a statesmanlike fashion.[10]

Back in Congress, it developed that the "9,000 acres" specified in the two Sparkman-Allen bills may have simply been a clerical error. Once the Forest Service produced an "official" map to be used in conjunction with S. 3224 (the bill containing wilderness designation), it appeared to duplicate the 11,000-acre proposal that the Alabama Conservancy had submitted to the Forest Service as the result of the meeting with the congressional delegation in the fall of 1971. Senator Jim Allen directed the Agriculture Committee staff to "make a note of that and that what we want to go by is the area shown on the map."[11]

Agriculture subcommittee hearings on the two Sparkman-Allen bills, and on S. 3699, the Talmadge-Aiken Wild Areas bill, were scheduled in Washington on July 20 and 21, 1972. Back in Alabama, the national forests had a new forest supervisor, Arthur D. "Dick" Woody, who immediately angered Sipsey supporters by scheduling a public hearing in the state on wild areas on the same day as the congressional hearings. "After violent protest," said Mary Burks, the Alabama hearing was rescheduled to July 17 in Tuscaloosa. Woody sheepishly claimed that he had learned of the scheduled congressional hearings "the following day" after sending out his first hearing notice.[12]

The testimony of Weesie Smith, who was now president of the Alabama Conservancy, pretty much sums up the conservationists' position at the Tuscaloosa meeting: "I feel that the Sipsey is eminently qualified for inclusion" in the National Wilderness System, but "if a new 'wild areas' classification will protect the unusual areas in our forest from ... destruction, I heartily endorse it."[13]

Shortly before the Senate Agriculture hearings, Sen. Henry Jackson introduced and had assigned to his Senate Interior Committee S. 3792, an "omnibus" eastern wilderness bill containing a 12,000-acre Sipsey Wilderness and proposed designations in eight other eastern and midwestern states. This served to underscore the jurisdictional dispute between the two committees, but it also reflected that the tide was beginning to turn against wild areas in favor of eastern wilderness. George Alderson, legislative director of Friends of the Earth, advanced an argument that influenced the Sierra Club to finally oppose the wild areas concept:

> The wild areas system would serve the Forest Service cause well. It sets two Senate committees to fighting over specific wilderness area proposals; it confirms the Forest Service "purity" argument; it fragments the wilderness movement into regional factions with less influence, instead of a unified national movement; it puts citizen environmentalists at a disadvantage in unfriendly committees of Congress. And it gives the Forest Service new hope for stopping the citizens' wilderness proposals in every western state.
>
> The Forest Service's objective on Capitol Hill is evidently to get eastern lands firmly away from the Interior Committee, where citizens have a great deal of influence, and let the Agriculture Committee do the dirty work of turning down all the wilderness or wild area proposals.[14]

Technically, the Sparkman-Allen bills and the Talmadge-Aiken bill were assigned to the Agriculture Committee's Subcommittee on Environment, Soil Conservation, and Forestry. The subcommittee's chairman, James Eastland of Mississippi, extended Jim Allen the courtesy of presiding over that part of the hearing that dealt with the two Alabama bills, and Senator Aiken, the hearing on his bill. The hearings covered two days, beginning July 20.

The witnesses were numerous; the hearing record, lengthy; and the testimony, all over the map.[15] Timber industry representatives opposed both eastern wilderness and wild areas. Perhaps the most concise expression of their position came from an Alabamian, Dirrill W. Moore of Sylacauga, who appeared in his capacity as president of the Southeastern Lumber Manufacturer's Association: "I support . . . the multiple use concept and I resent being excluded by any means or any purpose."

The Wilderness Society supported eastern wilderness, period. The Izaak Walton League maintained that areas in the South and East did not qualify for statutory wilderness designation and thus supported a system of wild areas, but with provisions improving the protections proposed by the Talmadge-Aiken bill.

Two witnesses appeared for the Sierra Club, J. William Futrell of Tuscaloosa, a member of the club's national board of directors, and Theodore Snyder Jr. of Greenville, S.C., a Sierra Club regional vice president. Reflecting the changing dynamics within the club, both men supported wilderness designation for the Sipsey, but both were open to a wild areas system, as well. Futrell testified against the Talmadge-Aiken approach, and instead suggested using as a guide S. 3225, the Sipsey Wild Area bill that had essentially been written by Mary Burks and Weesie Smith.

Mary, Weesie, and Charles Prigmore testified for the Conservancy. Briefly stated, they wanted no recreation area and a wilderness designation for at least 11,000 acres of the Sipsey. If, however, Congress chose the wild area route, then they too preferred the provisions of S. 3225 and requested that the entire 15,000-acre Bankhead Wilderness Study Area receive that designation.

The only member of the Alabama House delegation to testify in person was Congressman John Buchanan of Birmingham, who supported wilderness designation for the Sipsey: "I can't quote Thomas Jefferson, but agree with him concerning men in government such as those in the Forest Service, that chaining them, binding them by the chains of law, is the best way to conserve or protect anything. I think there is clearly need of congressional action in this case if there is to be protection for what we feel to be a very precious area."

The other Alabama congressmen all submitted written statements favoring protection of the Sipsey in one way or another, but Rep. Bob Jones's submission and the Nixon administration's response to it are worth dwelling on, as they address the National Recreation Area idea, one that Sen. Howell Heflin would resurrect when the Sipsey was proposed for enlargement in the 1980s.

Jones sought to justify his recreation area as a means to protect the Bee Branch canyons, contending that recreational use could be restricted there and diverted instead to other parts of the river's watershed: "Other canyons could be set aside for hiking and back-pack camping. Attractive sites could be prepared for those who use campers and trailers. A lodge and other facilities could be provided as required. A visitor center, along with trails identifying flora, could be a major nature educational facility. Parking areas, waste-disposal systems, and other facilities could be developed to service man's needs and lessen his impact upon nature."

The administration wasn't buying. J. Phil Campbell, undersecretary of agriculture, testified:

> The [Sipsey] Area which would be designated [wilderness] by S. 3224—except for about 400 acres in the Bee Branch Scenic Area—does not meet the criteria for wil-

derness set forth in the Wilderness Act... We also feel that the area is lacking in unique characteristics and the national significance required of a national recreation area. The character of the area is such that constructed facilities would have to be the major attraction with the surrounding land base serving as a secondary attraction. If a National Recreation Area were established by Congress, considerable expenditure of public funds would be required to develop and administer the area, since the area probably would not attract private investment capital.... [T]his Department strongly recommends that S. 3224 not be enacted.

Campbell also stated that the department and the Forest Service were leaning toward the wild areas concept, but they were still working on details and thus opposed enactment of both the Sparkman-Allen and Talmadge-Aiken Wild Areas bills.

Senator Allen concluded his portion of the hearings by addressing the Alabama Conservancy delegation:

Well, I wish to express my very deep appreciation to all of you and to all of those that you represent, for your dedication to this work. It is quite obvious to me that it is a labor of love on the part of all of you.... We certainly feel that it is the duty of Congress to listen to public opinion. We feel that you have made a great contribution to the concept of creating a wilderness area in this region, and we hope that in the not too distant future your work will fulfill your dreams. You picked a wonderful dream to work toward and I want to express the hope that all of your dreams be as lofty as this and that all of your dreams come true.

Senators Aiken and Talmadge, however, had a different dream, arguably a well-intentioned one, and Sparkman and Allen decided that the most practical thing was to go along. On September 11, 1972, the four senators cosponsored S. 3973, a new version of a National Forest Wild Areas Act, which established a 12,000-acre "Sipsey Wild Area within the Bankhead National Forest, Alabama, as a prototype." The Senate Agriculture Committee favorably "reported out" (that is, recommended to the full Senate) the bill on September 22, saying, "It would be counter-productive to soften the regulations of the Wilderness Act but, for a variety of reasons, these areas of the Eastern forests must receive protection similar to that of the wilderness areas. This is the purpose of the bill."[16]

The new legislation was similar to the earlier Talmadge-Aiken bill, but with strengthening provisions insisted upon by the Alabama Conservancy and Senators Allen and Sparkman:

Prohibiting in the Sipsey Wild Area
  a) commercial enterprises;
  b) permanent roads and, except for administrative purposes and to protect public health and safety, any temporary road;
  c) use of mother vehicles, motorboats or aircraft; and
  d) any structure or installation.

The National Forest Wild Areas Act of 1972 passed the U.S. Senate by voice vote (with no debate, little prior notice, and few Senators in attendance) on September 26, 1972. Dennis Roth says the Wilderness Society was "caught off guard" by Senate passage, but its congressional supporters were able to kill the measure in the Agriculture Committee of the House. The 92nd Congress then permanently adjourned.

In December 1972 the Wilderness Society invited eastern leaders to a conference in Knoxville, Tennessee, to plan a 1973 campaign for a new wilderness bill patterned on the one that had been introduced by Senator Jackson back in the summer. Capping a year that she now characterizes as "a seething mass of happenings," Mary Burks attended on behalf of the Alabama Conservancy. The tone of the statement she made there reflects the year's exhausting work, the constant travel, the emotional highs and lows, and, finally, resignation toward what the Conservancy and its board of directors understood to be the political reality of the time:

> The Wild Areas Bill that finally emerged, due to Conservancy insistence, incorporated every safeguard of the Wilderness Bill. . . . [We have] the unanimous support of the entire Alabama delegation and their commitment to work for early passage of Sipsey legislation in 1973.
>
> We cannot ask our delegation to completely change direction. We cannot ask our Southern Senators, one of whom is a member of the Agriculture Committee, to remove our bill to the Interior Committee which is dominated by Western Senators interested in Western Wilderness. We must face the realities of political life in Alabama. We cannot and will not commit ourselves to destroy the Eastern Wild Areas Bill, thereby destroying the Sipsey for which we have struggled unceasingly since our organization was founded.[17]

CHAPTER FIVE

A Truly National System

As 1972 CAME TO A CLOSE, Sipsey supporters found that their little wilderness had been the centerpiece of six separate bills in Congress: the Jones-Bevill National Recreation Area bill, the initial Sparkman-Allen Wilderness bill, their subsequent wilderness–recreation area and wild area bills, Senator Henry Jackson's "omnibus" eastern wilderness bill and, finally, the Talmadge-Aiken Wild Areas bill. None of them had passed the 92nd Congress by the time it adjourned, so the process would begin once more in 1973, with a two-year window of opportunity afforded by the 93rd Congress.

It was far past time for the national conservation groups to decide what they were going to support. Following its 1972 annual meeting, the Izaak Walton League came out in favor of eastern wilderness over wild areas, leaving only the Sierra Club as the major holdout. The meeting in Knoxville that Mary Burks attended at the end of 1972,

was the first step in overcoming Sierra Club resistance to eastern wilderness legislation. The Wilderness Society also formed a very successful "front" group called "Citizens for Eastern Wilderness," the main purpose of which was to be a face-saving device for Sierra Club members who wanted to discreetly align themselves with the society. . . . The Sierra Club eventually endorsed the society's position but not before there had been "difficult" meetings and some bad feelings between officials of the two organizations.[1]

Theodore Snyder of Greenville, S.C., a Sierra Club regional vice president and Wilderness Committee member, sponsored the resolution endorsing eastern wilderness that the club ultimately passed. Snyder says Peter Borelli, the Sierra Club staff person assigned to handle the southeastern proposals, including the Sipsey, had previously convinced club leaders "that we could not get

Wilderness in the East; that it was impossible, and to forget it. Based on Peter's strong advice, we gave up on Wilderness." However, after "long and fateful argument with Ernie Dickerman" of the Wilderness Society, Snyder was convinced to approach the Sierra Club board with a resolution in favor of eastern wilderness. "With that authority," says Ted Snyder, "we commenced lobbying in earnest."[2]

Meanwhile, Sipsey supporters in Alabama were confronting a dilemma of their own. On January 4, 1973, Senators Aiken, Talmadge, Allen, and Sparkman, now joined by Hubert Humphrey of Minnesota, introduced a new version of wild areas legislation, S. 22, again with a 12,000-acre Sipsey Wild Area as a prototype. The previous day, Alabama Congressmen Jones, Bevill, and Buchanan introduced H.R. 656, which they intended as companion legislation to the wild areas bill in the Senate, again with a 12,000-acre Sipsey prototype. As before, these bills were assigned to the respective Agriculture Committees.

By now the sentiment for eastern wilderness was gaining serious momentum. On January 11, 1973, Sen. Henry Jackson and eighteen other senators, many of whom had been sponsors of the original Wilderness Act, introduced S. 316, which became popularly known as the Eastern Wilderness Areas Act, designating twenty-eight new wilderness areas in sixteen states, including a 12,000-acre Sipsey Wilderness at the top of the list. Companion House legislation, H.R. 1758, was sponsored by Pennsylvania congressman John P. Saylor and others the same day. Subsequently, Birmingham representative John Buchanan signed on as a cosponsor. These bills were assigned to the respective Interior Committees, renewing the jurisdictional conflict with Agriculture.

The sponsors of the two eastern wilderness bills had a message for the U.S. Forest Service. Said Senator Jackson:

> A serious and fundamental misinterpretation of the Wilderness Act has recently gained some credence, thus creating a real danger to the objective of securing a truly national wilderness preservation system. It is my hope to correct this false so-called 'purity theory' which threatens the strength and broad application of the Wilderness Act. . . . I remind my colleagues again that a central purpose of the Wilderness Act of 1964 was to reserve to the Congress the authority for determining what areas could be designated as wilderness. It is not up to an administrative agency to make this decision.[3]

The House sponsor, John Saylor, was more confrontational:

If [eastern wilderness opponents] want to come before me with a lot of hokum about "purity" and "diluting the high standards of the Wilderness Act" and so forth, they are welcome to do so, but I ask them to come with their eyes opened and prepared for battle. . . . If the U.S. Forest Service or its officials attempt to subvert the Wilderness Act and the national wilderness preservation system, or usurp the powers of the U.S. Congress in this field, they had best be prepared for a monumental struggle with this House.[4]

The battle lines as drawn left the Sipsey Wilderness Area stranded alone between the opposing forces. Mary Burks summarized the options to a January meeting of the Alabama Conservancy's board of directors:

Support Eastern Wilderness Bill and withdraw from Wild Areas Bill.

This would probably alienate Senator Allen who has staunchly supported Sipsey from the beginning. . . . He is quite proud of Wild Areas and believes it is better than Wilderness Act.

Allen's legislative assistant wants us to support Wild Areas. He would not say we should withdraw from Jackson Bill but stated that eventually we would have to choose when giving our testimony before a committee.

Remain in both bills.

The Wilderness Society [is] . . . very anxious for us to do this as they are very much disturbed over fighting the Sipsey when they oppose the Eastern Wild Areas System.

Sierra's [staff member Peter] Borelli says that if we go this route Sipsey will be the only area in both bills. This is true. Every other eastern area which has been proposed for Wilderness has been withdrawn from Wild Areas Bill by its [sponsor]. . . . We alone are to be made an example of.

Representative Buchanan's legislative assistant suggested that we remain in the Wilderness Bills but be very low key in our support. If Wild Areas should be defeated, this would leave us with an alternative. Otherwise the Sipsey goes [down] with the Wild Areas Bill.[5]

"How did we ever get into this mess?" lamented Conservancy president Weesie Smith to the press. "You wouldn't believe this thing. We're in the middle and whatever we decide will be wrong."[6]

Actually, there was really only one course to follow. The board voted to remain in both proposals. "Congress will have to decide which is best," said Weesie. "It is their decision to make anyway."[7]

The momentum for eastern wilderness was further strengthened when President Nixon more clearly endorsed the concept in his 1973 environmental message to Congress. Faced with a directive from the president, the Forest Service decided to drop the idea of a separate wild areas system. However, they couldn't bring themselves to abandon the purity standard altogether, as it was crucial for denying new wilderness proposals in the vast western national forests. Thus the agency came forward with another tactic, and this one contained a surprise for Sipsey backers.

S. 316, Senator Jackson's Eastern Wilderness Areas Act, was set for hearing in Washington on February 21, 1973, before the Public Lands Subcommittee of the Senate Interior and Insular Affairs Committee. There, Forest Service chief John R. McGuire proposed, instead, the Eastern Wilderness Amendments Act of 1973.[8] It directed the secretary of agriculture to "review" fifty-three specific Wilderness Study Areas in the East to determine their "suitability or nonsuitability for designation as additions to the national wilderness preservation system." Heading the list, in now a fifth piece of congressional legislation in just two months, was a Sipsey Wilderness Study Area, but of only 9,400 acres, confounding supporters in Alabama.

Besides failing to designate any "instant" eastern wilderness candidates, forcing them instead back into the clutches of the Forest Service for "review," McGuire's proposal also sought to amend the Wilderness Act itself by creating two categories of wilderness, one in the East and one in the West:

> [W]e urge that a distinction be made between National Forest lands located in the east and those areas west of the one hundredth meridian. . . . [Under] the draft legislation we have transmitted . . . only in National Forest System units east of the one hundredth meridian the Secretary could consider for review areas once significantly affected by man's works, but which have now become areas of land where the imprint of man's work is substantially erased [and] which have generally reverted to a natural appearance. . . . [W]e recommend that the western areas be preserved with the primary objective of identifying and protecting a unique, basically undisturbed, primeval, non-renewable natural resource.[9]

McGuire received no encouragement from the members of the subcommittee:

Senator Church: The adoption of the amendment you propose setting up a different criteria for the eastern area, would in fact confirm the purity train, so-called, that

you have been applying as a test for the eligibility of given areas to became a part of the wilderness; would it not?
Mr. McGuire: Yes; I suppose it would. . . .
Senator Haskell: I think the cat is now out of the bag. I couldn't understand . . . how you could possibly interpret your definition of wilderness the way you do. What I gather now is that you are afraid all of the area that qualifies under the definition will be designated as wilderness areas. . . . [I]t is up to Congress to decide which of those areas will be included. You seem to fear that just because an area meets the definition it will be included. . . .
Senator Church: Bring your proposals up here; Congress is the final judge of what goes into wilderness and what does not.[10]

The conservation organizations finally presented a united front at the hearing. With only some hedging here and there—the National Wildlife Federation, for example, preferred the two categories of east and west proposed by the Forest Service—all the national conservation groups, Sierra Club and Izaak Walton League included, endorsed the concept of eastern wilderness as proposed by S. 316. And Mary Burks testified in favor of the bill and a 12,000-acre Sipsey Wilderness.

The source of the reduced acreage for the Sipsey in the Forest Service proposal proved to be the Alabama forest supervisor's office. The agency wanted to exclude large tracts along the southern and northwestern boundaries, on the justification that there were some private inholdings on the periphery and "three cemeteries, several of which have recent interments," and a road to one of the cemeteries. Besides, said the Forest Service, the excluded acreage "contains none of the canyon area and is not needed as a buffer, and contributes nothing to the wilderness experience not already found" in the canyons.[11]

These were lame justifications. The private land, being on the periphery, could easily be gerrymandered out by following the public land lines (which is exactly what was ultimately done). Families of those buried in the cemeteries retained the right of access to them under the Wilderness Act, so that even though the road was eventually gated, the families could still use it to reach the burial plots (and do so today). The true intent of the Alabama forest supervisor's proposal was to limit the wilderness to the canyons so that parts of the uplands could continue to be logged.

The Forest Service's acreage reductions would prove to be a serious problem for Sipsey supporters. In the spring of 1973 the chairmen of the Senate Interior and Agriculture Committees agreed to end their dispute and to share jurisdiction over all eastern wilderness legislation. The plan was for the Interior Committee to finish its work on S. 316, the Eastern Wilderness Areas Act, then refer it to Agriculture for consideration. There, the attitudes of Chairman Herman Talmadge could have a significant impact on proposals from senators all over the country, and especially upon fellow committee member Jim Allen of Alabama. Talmadge wrote to another senator, "I [believe] that S. 316 gives too much power to the Congress. Wilderness designation ought to spring from professionals. . . . It is my understanding that the Forest Service is generally satisfied with the designations of instant and study areas in S. 316, but if, as I understand, this bill will be referred to my committee, I will have to rely heavily on the advice of the Forest Service insofar as areas included and the acreage therein."[12]

As time went on, the pressure increased on the Alabama Conservancy to accept the Forest Service's reduced acreage so that Jim Allen would not be forced to confront Chairman Talmadge. The result was a rift between Mary Burks and Charles Prigmore. Mary was adamant about holding out for the 12,000 acres, while Prigmore was willing to settle for smaller acreage as a compromise. "They were both motivated by a desire to succeed, and to do what was right. But Charles was a good compromiser. He felt that we needed to get something on the books," says Elberta Reid. "Mary and Charles really got into an altercation over this." In the end, the Conservancy decided to support Mary and to continue to push for 12,000 acres.

In June 1973 a new element was introduced into the mix of Sipsey preservation initiatives. Back in 1968 Congress had created a National System of Wild and Scenic Rivers, just as it had created the wilderness system four years before: "It is hereby declared to be the policy of the United States that selected rivers of the Nation which, with their immediate environments, possess outstandingly remarkable scenic, recreational, geologic, fish and wildlife, historic, cultural or similar values, shall be preserved in free-flowing condition, and that they and their immediate environments shall be protected for the benefit and enjoyment of present and future generations."[13]

Reflecting the continued in-state popularity of the Sipsey cause (and prodding by the Alabama Conservancy), on June 13, Congressmen Jones and Bevill sponsored H.R. 8463, to designate the West Fork Sipsey for official study for inclusion in the National Wild and Scenic Rivers System. Shortly thereafter, Senators Sparkman and Allen introduced companion Senate legislation, S. 2216. Although this proposal would not be a factor in the initial establishment of the wilderness itself, it would have a significant impact on the success of its enlargement in the 1980s.

For now, however, the problem was to get anything at all passed by Congress. By December 1973 both the Senate Interior Committee and the Senate Agriculture Committee had reported out their respective versions of the Eastern Wilderness Areas Act and the Wild Areas System Act, but the Congress and many of its leaders were now becoming preoccupied with the Watergate scandal, the possible impeachment of Richard Nixon, and all the implications for pending legislation that the situation presented. Moreover, Senator Jackson and others in the Senate Interior Committee had requested a "hold" on the wild areas bill, aggravating its sponsors. Mary Burks reported to her board of directors that "Senator Allen has been very vague in his promises. [Charles] Mitchell, his legislative assistant has been positively unpleasant. They are bitter over . . . the hold placed on" the wild areas bill. Then, the death of Pennsylvania congressman John Saylor, the principal House sponsor of the Eastern Wilderness Areas Act, brought the progress of his bill in that body to a halt.[14]

Confounding matters for Sipsey supporters was that, at the last minute, the Senate Interior Committee acceded to the wishes of the Forest Service and reduced the Sipsey acreage to 10,600 in the final version of S. 316.

Nevertheless, 1974 saw the prospects of the eastern wilderness movement still on the rise. For several months, Senator Aiken had been expressing a willingness to consider approaches other than his wild areas bill: " 'I don't care what these areas are called or whose name is on the legislation that is enacted as long as we can protect certain areas in their native wild state for future generations to enjoy.' The Vermont Senator . . . pledged that he is ready to work with others to insure that some legislation can be passed by the 93rd Congress to create a wilderness or wild areas system in the East."[15]

Mary Burks got on well with Senator Aiken and she believes she may have had some influence on his thinking with respect to wild areas. The Sipsey was the only specific proposal for "instant" designation still left in the Wild Areas Act. Mary went to Aiken and asked him to withdraw the Sipsey from his bill and allow it to proceed as an eastern wilderness proposal. "He said, 'I had no idea I was causing any distress. I was just trying to help,' " recalls Mary.

As previously agreed between the committee chairmen, S. 316, the Eastern Wilderness Areas bill, was referred to Agriculture.

The Chairmen of both committees instructed their staffs to work together and attempt to draft a compromise bill. . . . After several working sessions, and several draft bills, the staffs of the two Committees were able . . . to reach agreement on recommendations for a proposed revised bill. Since Chairman Talmadge and Senator Aiken believed that it was not important whether the areas to be preserved are called wild areas or wilderness areas, the revised bill followed the philosophy of S. 316—the extension of the National Wilderness Preservation System established by the Wilderness Act of 1964. Once the decision was made to extend the existing system, the Chairman and Senator Aiken felt that it would be wise to make management in the National Wilderness Preservation System uniform, both east and west of the 100th meridian. . . . On April 24 [1974], the Committee on Agriculture and Forestry . . . considered the major issues involved and ordered reported an original bill, combining, in their judgment, the best features of S. 22 [the Wild Areas Act] and S. 316.[16]

The result was the official end of wild areas, once and for all, and the birth of S. 3433, the Eastern Wilderness Areas Act of 1974. It contained a restored 12,000-acre Sipsey Wilderness proposal.

Jim Allen wrote to Alabama Conservancy board member Lyle Taylor of Huntsville,

[P]ermit me to advise that earlier today our Senate Agriculture Committee . . . adopted my amendment to increase the proposed Sipsey Wilderness to 12,000 acres instead of the 10,600 acres as called for by the Senate Interior and Insular Affairs Committee in . . . its version of the legislation.

It is my considered opinion that we have approved a Wilderness bill which is much better than S. 316. . . . My prime concern for the past five years has, of course, been the preservation of the Sipsey/Bee Branch area of the Bankhead National Forest, as evidenced by the several bills Senator John Sparkman and I have introduced in the 91st, 92nd and 93rd Congresses. I shall continue to do all I can to make this a reality.[17]

Meanwhile, new House sponsors had been recruited for the Eastern Wilderness Areas Act, and legislation designated as H.R. 13455 was introduced on March 13 by the Democratic chairman and subcommittee chairman, and the respective ranking minority (Republican) members, of the appropriate House Interior Committee bodies. This legislation generally tracked S. 316 and would need to be reconciled with the new Senate bill—or vice-versa.

On May 31, 1974, the U.S. Senate passed the Eastern Wilderness Areas Act of 1974, as written by the Agriculture Committee.

[Mary Burks and Weesie Smith], both of Birmingham, and Dr. Charles Prigmore of the University of Alabama, all past presidents of the [Alabama] conservancy, were singled out for praise on the Senate floor . . . by Sen. Jim Allen who has worked for passage of the bill.

"It's the most exciting thing that's happened to us in years," said Mrs. Mary Burks. . . . "We have great hopes we can expedite it in the House," she said, expressing the fears of conservationists that the bill's passage may be slowed by impeachment proceedings [against President Nixon] on the House floor.[18]

But not even the possible impeachment of a president could stop the eastern wilderness juggernaut, although it did force the final resolution to be delayed until the waning hours of the 93rd Congress.

Although the House version contained a full 12,000-acre Sipsey Wilderness, it was generally considered to be "weaker" than the Senate bill on such technical issues as treatment of outstanding mineral rights in areas to be designated, none of which were particularly significant to the Alabama proposal. Additionally, Congressman John Melcher of Montana, chairman of the House Public Lands Subcommittee, took the unusual step of requiring written consent from every congressman in whose district a proposed wilderness or wilderness study area was situated. Dennis Roth says, "This procedure had the effect of halving the number of acres that were to go into the system immediately or that the Forest Service was to study for possible future designation."[19] Fortunately, this was not a problem for the Sipsey, for by now all Alabama's congressmen had signed on as cosponsors of the House bill.

But it was a problem for the Wilderness Society and local groups that had wilderness proposals removed from the legislation, a problem that grew more acute when the House passed the legislation on December 18, 1974. With only a few days left in the 93rd Congress, not enough time for a formal House-Senate Conference to reconcile the differences with the Senate bill, and facing the prospect of having to start entirely from scratch in the 94th Congress, the Wilderness Society's Ernie Dickerman notified Sen. Henry Jackson that his organization would accept the House bill. The Senate passed the House bill on December 19, and Pres. Gerald Ford signed the new legislation on January 3, 1975.[20] The long battle to achieve recognition and protection of eastern wilderness was over. And Alabama had itself a dandy little national wilderness area.

On January 4, 1975, President Ford also signed the legislation designating the West Fork Sipsey as a study river for the National Wild and Scenic Rivers System.[21] The study was assigned to the U.S. Forest Service, with a directive that it complete its work and submit its recommendations to Congress by October 2, 1979. In fact, however, the draft recommendation was not released until the spring of 1984, and, as it turned out, the delay was a propitious one.

Before filing his agency's final "official" map with the Congress to accompany the passed legislation, Alabama forest supervisor Dick Woody actually increased the size of the Sipsey Wilderness Area to about 12,700 acres. The additional acreage, he said, had the purpose of creating more manageable boundaries. The new wilderness lay within an oval surrounded by roads known as the Cranal Road, the Kinlock Road, the Norwest Road and the Bunyan Hill Road, with the eastern boundary being the easterly escarpment of Borden Creek (see map p. 2). Private lands just to the west of Borden Creek and in the southwest corner and along the western periphery of the oval were excluded, but a couple of forty-acre tracts deep within the wilderness were still in private hands.

Condemnation was one of the national issues that had to be resolved before the eastern wilderness legislation could pass. It was not a factor as far as the Sipsey Wilderness was concerned, but supporters of other eastern proposals felt the need to grant the power to the Forest Service in order to protect the integrity of their areas. Thus while the original 1964 Wilderness Act contains no power of eminent domain, the 1975 legislation does. Briefly stated, the government was granted that authority in the event private owners used their property in a manner inconsistent with the Wilderness Act and failed after notice "to promptly discontinue such incompatible use." Accommodation was made, however, for owners occupying their property for residential or agricultural purposes, allowing them (or their heirs) to continue that use for twenty-five years or the term of their natural lives, whichever came later.

The condemnation power was, in fact, the only difference between the original 1964 Wilderness Act and the 1975 bill. People in national conservation groups who were intimately involved with the 1975 legislation are quick to correct anyone who calls it the Eastern Wilderness Act, as the entire focus of the movement was to end the distinction between east and west, discredit the purity standard, and create a truly national wilderness system. Douglas W. Scott, formerly with the Wilderness Society, points out that the act itself has no title, but recites simply that its object was "[t]o further the purposes of the Wilderness Act." The correct reference, says Scott, is to use the name by which it was popularly known at the time, the "Eastern Wilderness *Areas* Act."[22]

Be this as it may, the condemnation power does significantly distinguish the two pieces of legislation, and its existence was widely used by wilderness opponents attempting to defeat new Alabama proposals in the late 1970s and 1980s.

CHAPTER SIX

Dedication for the Future

A FORMAL DEDICATION CEREMONY for the Sipsey Wilderness was held on May 17, 1975, with more than two hundred people in attendance at a site just above the East Bee Branch Canyon soon to be closed to vehicular access by wilderness management. The Birmingham Audubon Society arranged for its annual spring picnic to coincide with the event and bused in a full load of folks. Congressman Tom Bevill delivered a few predictable platitudes. The dedication itself was conducted by the executive director of the Wilderness Society, Stewart Brandborg of Washington, D.C., but it was the writer Mike Frome, author of *Battle for the Wilderness* and *Conscience of a Conservationist*, who precisely defined what the Sipsey Wilderness movement had been all about: "This wilderness area is a 'reaffirmation of man's faith in himself and it's proof that the people, and not the bureaucrats, will have the final say.'"[1]

One of the bureaucrats in attendance was Bankhead district ranger William J. "Bill" Bustin, who had been assigned to assist Mary Burks with the logistics for the ceremony. Mary found Bustin to be "ambivalent" about the wilderness campaign and "helpful" when requested, which was a far cry from the way he would behave when the Sipsey was proposed for enlargement in the 1980s.

But that was for the future, when a different dynamic would hold sway. In the 1970s bureaucrats within the national forests in Alabama had been taken completely by surprise by the groundswell of popular support for preserving the Sipsey, as well as by the rapid political acceptance the effort gained in Alabama's congressional delegation. Since the agency's national office was itself struggling, and failing, to develop a coherent policy on eastern wilderness, the Forest Service in Alabama could do little to influence events.

Much the same was true regarding the commercial timber industry. Although its national organizations made desultory appearances at various hearings to oppose wilderness designations in the East, it, too, was stinging from popular outcry against the practice of clear-cutting the public lands. In Alabama, timber industry opposition to a Sipsey Wilderness would have meant bucking the wishes of both of the state's U.S. senators and its entire House delegation; thus it never materialized.

The Sipsey Wilderness campaign, begun as a purely local effort, had been swept into a great push to force change in national policy through the power of elected officials in the Congress. After so much early work and exciting progress in channeling public support for Sipsey preservation, Alabama activists found their issue essentially taken away from them and their role transformed into one of weary perseverance. David Saylor, Washington coordinator for Citizens for Eastern Wilderness, put it this way: "The Congressional game is largely one of attrition. The winners are those who stick in there relentlessly, endlessly pushing through the dull periods, and in the end triumphing because of their frequent and repeated application of pressure until the final victory is won."[2]

The impact of the Eastern Wilderness Areas Act was of truly historic proportions, not just in terms of the acreage "instantly" protected, but also in its portent for the future. The years to come would see state after eastern state successfully preserve threatened wilderness areas on their public lands. In 1984 alone eight southern states enacted 352,855 acres of national forest wilderness (Texas, 34,400 acres; Arkansas, 91,000 acres; Mississippi, 5,500 acres; Tennessee, 24,942 acres; Georgia, 14,529 acres; Florida, 49,150 acres; North Carolina 68,750 acres; and Virginia, 64,584 acres). Alabama, unfortunately, was not among them, as local wilderness opponents both in government and in industry determined that the issue would not pass them by default again.

Future wilderness opposition in Alabama was fueled in some measure by the naïveté of early Sipsey supporters. Faced with opposition at all levels of the U.S. Forest Service, the Alabama Conservancy believed its only hope was to severely limit what it asked for. "We really didn't know what we were doing," says Mary Burks. "We had no idea we could propose road closings or protection for entire watersheds. If we knew then what we know now, we would have asked for three times as much."

To Sipsey leaders, getting any wilderness established at all in an Alabama forest was such a daunting task that the notion of a bigger Sipsey preserve—or, for that matter, designated wilderness in any other Alabama national forest—seemed as remote as the moon. Thus it was understandable that they often told Forest Service officials and congressional staffers that the Sipsey would be Alabama's only wilderness proposal. The Forest Service's Bill Bustin would remember that, as would Charles Mitchell, Senator Allen's legislative assistant, and together they would use it to influence the politicians who succeeded to Alabama's Senate seats in the 1980s—Jeremiah Denton and Howell Heflin.

Nevertheless, the original Sipsey Wilderness had been restricted only to the climax, if you will, of the West Fork Sipsey watershed, where the river and its most pristine side canyons made the deepest canyon cuts and harbored the least evidence of the hand of man. Most of the river's upper watershed, its primary tributaries and the source of its life, were left out. When the opportunity came in the 1980s to rectify the omission, none of the leaders of the new preservation campaigns felt bound by the representations of their predecessors. No one could have foreseen that an obscure southern governor named Jimmy Carter would ascend to the presidency of the United States and would institute something called RARE II, leading to a tripling of Alabama's national forest wilderness acreage by the end of the century.

PART TWO

The Cheaha Wilderness Area

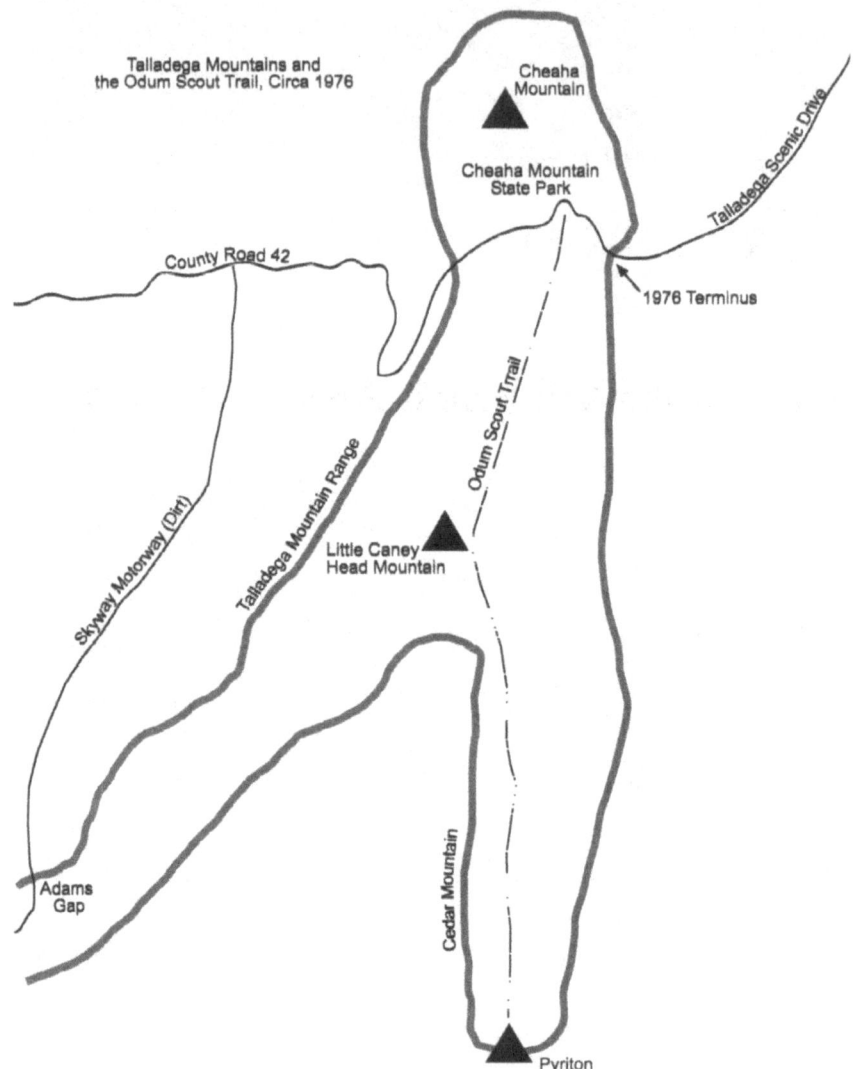

Talladega Mountains and the Odum Scout Trail, Circa 1976

CHAPTER SEVEN

Nobody Here Felt Like It Was Hurting Anything

MY OWN FORAY INTO WILDERNESS preservation began not in the Sipsey, which was to later consume my existence, but in the mountains of the Talladega National Forest. There, in 1975, I entered in pursuit of natural beauty and emerged, unexpectedly, with a cause.

At the time, the Talladegas were the only mountains in Alabama both in public ownership and open to use by the general population. Shoving against the geological formation known as the Ridge and Valley Province, this single range is the southern extremity of the Piedmont Plateau as well as the true terminus of the Appalachian Mountains. Although gold was discovered here in the early 1900s, and was even produced in some marketable quantities before playing out, the essential ruggedness of the range discouraged habitation and development, rendering it a logical target for acquisition when the Talladega National Forest was established in 1936.

About two-thirds up the range lies Cheaha Mountain, Alabama's highest point and the site of the popular Cheaha State Park. Immediately to the south is a roadless mountain backcountry of some 12,000 acres, encompassing the ridgeline called Talladega Mountain. Here, in the 1960s, local Boy Scout troops developed a hiking trail running south from Cheaha State Park, down the top of the ridgeline, through the entire length of the roadless area, a distance of some ten miles. The Odum Scout Trail, as it was called, quickly became known in scouting circles. Situated roughly half-way between Birmingham and Atlanta, it drew hikers from both metropolitan areas. By 1978 it was receiving regional use; an estimated thirty thousand boys from all over the South had been awarded Scout patches for completing the trek.[1] It was far and away Alabama's best-known hiking trail.

My exploration of the high Talladegas led me to a special cliff affording westerly vistas, across the Coosa River valley to the hills called the Sleeping Giants, standing sentinel over the city of Talladega, and beyond, on a clear day, to the twin Oak Mountain ridges that cradle metropolitan Birmingham on the southeast. To the north, just hidden by the broad shoulder of Cheaha Mountain, lay the towns of Oxford and Anniston, where nighttime artillery practice at the U.S. Army's Fort McClellan lit up a camper's sky in spectacular display. Beyond lay Lookout Mountain and the great Sand Mountain Plateau dividing the Coosa and Tennessee Rivers, all visible from my special cliff.

The U.S. Forest Service had big plans for the area, however. My first inkling came in a 1975 visit, when I encountered a line of survey markers running in ominous straightness along and beside the Odum Scout Trail, mile after mile. It looked to me like a highway route; why else would anyone spend the kind of money required for surveying a wilderness mountain ridge? And a highway is exactly what it proved to be.

The Talladega Scenic Drive was the brain child of the U.S. Forest Service's Alabama office, specifically of one James E. "Jim" Bylsma, assistant forest supervisor in charge of recreation planning for the national forests in Alabama. "I've been associated with this for years, and have been promoting it," he told the press.[2] Planning began in the 1950s. The idea was to target the Talladega Mountains for "intensive recreation development," accessed by a paved highway, constructed at public expense, running along the ridge tops the entire seventy-four-mile length of the Talladega National Forest. In so doing, it would bisect two popular public hunting areas, the Choccolocco Wildlife Management Area in the north, and the Hollins Wildlife Management Area in the south, and run smack down much of the Odum Scout Trail.

Jim Bylsma's search for support for the highway contributed to the formation of a new East Alabama tourism group, the TallaCoosa Highland Lakes Association, which promoted the drive as a magnet for tourists in "Alabama's Smokies."

Local officials were sold on the idea, and Congressman William F. "Bill" Nichols, Democrat of Sylacauga, was asked to locate federal funding for construction, while the Forest Service acquired title to the little remaining private land lying along the right of way. The drive became a personal project of Congressman Nichols. U.S. senator Jim Allen was also a supporter.

I was appalled. The notion that the highest and best use of these publicly owned mountains was a highway, one which would destroy the Odum Trail and send tourist automobiles into public hunting areas, was completely alien to me. Until confronted with this, I was a conservative young business lawyer, enjoying the outdoors, "consuming" them, if you will, but doing nothing to protect them. The Talladega Scenic Drive turned me into an activist and forever altered my life.

Had there been a Delta Force Rangers against Roads, a Commandos for Conservation, I would have signed right up; as it was, I joined several local environmental groups and began to talk up opposing the scenic drive. A ready ally was found in J. Walden "Wally" Retan, recently elected as chairman of the Cahaba Group of the Sierra Club, located in Birmingham. In his precise, clipped way of speaking, Wally said, "I think this is something we should look into."

Wally's energetic attack produced a Sierra Club letter-writing campaign to the state Highway Department and the U.S. Forest Service, challenging the need for the road, and attracted citified media attention toward what had been a sweetheart development for east Alabama officialdom. The public relations officer of the Alabama Highway Department complained to the *Birmingham News*, "We've had a lot of trouble with environmentalists saying it [the scenic drive] will destroy the beauty. But that's just a small group and nobody here felt like it was hurting anything. We don't initiate many roads that aren't wanted. Someone had to have asked for the scenic drive."[3]

As entertaining as bedeviling the Highway Department might be, any serious effort to stop the road had to begin with the agency that conceived the plan and controlled the right of way, the U.S. Forest Service. In the summer of 1976 Wally Retan and I met with Alabama forest supervisor Dick Woody and grilled him about the scenic drive.

Yes, said Woody, this project was of long standing, but no final decisions had been made. Of course there would be proper environmental studies done before any further construction. No, despite the survey work along the Odum Trail, the route south of Cheaha Mountain had not been finally decided. There would be public hearings, and he would be sure to notify us when they were scheduled.

What Forest Supervisor Woody did not tell us was that bids were at that very moment being solicited for construction of a new segment of the highway, and that, since funds to build the drive into the Odum Trail backcountry had already been earmarked, planning for that section was being rapidly concluded. The fate, therefore, of the popular Scout trail and its protective wilderness might have been quite grim had the Talladega Scenic Drive planners not been caught with their legal trousers around their ankles.

The National Environmental Policy Act (NEPA) was designed to force federal bureaucrats who control public monies to give some thought in advance to the environmental consequences of the projects they adopt.[4] Except where mandated by other laws—the Endangered Species Act, for example—NEPA does not require that the government necessarily abandon a project because of adverse consequences; only that it consider mitigating alternatives, and do so in public, so that balancing forces may come into play. The act has gained general acceptance since its passage not only because worthy projects have not been hindered but also because it has helped improve and gain public acceptance for many others.

NEPA's basic requirement is that before embarking upon a "major federal action significantly affecting the quality of the human environment," the government must prepare an environmental impact statement (EIS), giving honest consideration to alternatives that might diminish adverse environmental consequences. By 1976 regulations and case law interpreting NEPA had established beyond question that a highway project could not be developed piecemeal without a comprehensive EIS addressing the consequences of the project as a whole. Talladega Scenic Drive developers had no intention of complying with this law.

Prior to NEPA's enactment, the first of three scenic drive segments designed to connect Cheaha State Park with I-20 was constructed. By 1976 an additional three-mile segment had been completed to the base of Cheaha Mountain, terminating at what is today the parking area for the Cheaha trail system. NEPA was by then the law of the land, and the feds had been careful to publish an environmental statement, since this segment pushed the road into Cheaha State Park, but the EIS covered only this three-mile stretch. The bids that were being sought at the time of our meeting with Forest Supervisor Dick Woody were, fortunately, not for the Odum Trail backcountry but for the last stretch crossing I-20 to the north. Since no significant public attention or controversy had heretofore been focused on the drive, and all of local officialdom was in happy concurrence, the contracts for this third segment were awarded with no prior environmental impact statement at all. The developers had no plan to produce a comprehensive EIS on the seventy-four-mile project or to consider any alternatives to the destruction of the Odum Scout Trail and the invasion of the two public hunting areas.

None of the state's environmental groups had any particular objection to improved access to Cheaha State Park. Indeed, the existing approaches from both the east and the west were torturously winding and dangerous in traffic. But if we were to challenge the road as a whole, we had to act immediately, before construction of the third segment was begun and the continuing expenditure of money tipped the "equities" of the situation to the developers. There was also something to be said for the tactical advantage to be gained by quickly shutting down the road in court.

The senior partner of my law firm, Samuel L. Tenenbaum, approved my taking the case on a reduced fee basis, but we needed plaintiffs, people and organizations that regularly used the Talladega Forest and would be adversely impacted by the highway. Wally Retan agreed to be one plaintiff; moreover, he underwrote the cost of the litigation until a Sierra Club fundraiser could kick in. But the Sierra Club itself, with its labyrinth of national rules and procedures required for approval of litigation, was not a practical candidate as a party. Thus we turned to the state's major independent environmental group, the Alabama Conservancy, creator of the Sipsey Wilderness Area. Its board of directors readily approved its participation.

On December 6, 1976, *The Alabama Conservancy and J. Walden Retan, as plaintiffs, vs. Federal Highway Administration, and others, as defendants* was filed in the U.S. District Court for the Northern District of Alabama. The suit sought an immediate injunction (called a temporary restraining order, or TRO) against construction of the third segment, and a final injunction against the Talladega Scenic Drive until a comprehensive EIS was performed on the entire seventy-four-mile length of the highway project.

A TRO can theoretically be obtained on the same date that an injunction suit is filed; indeed, it is proper procedure to ask for one, although even if granted, it is usually effective for only a few days, until the case can be presented in open court for a preliminary injunction that will last for the life of the litigation. Thus, after filing the suit, Wally Retan and I presented ourselves to the chambers of presiding U.S. district judge Frank McFadden to seek our restraining order.

I introduced myself to the judge's secretary as the attorney for the plaintiffs and Wally as one of the complaining parties. Her quizzical look took in my bearded face and shaggy hair, and Wally's conservative, well-groomed appearance. "I would have thought it was the other way around," she said.

The government was represented by Assistant U.S. Attorney Henry Froshin. Though able and experienced, Froshin had difficulty defending his clients' position. He could argue only that the highway was being built in an "environmentally sound" manner, with grassy shoulders and scenic overlooks. "But, Henry," said Judge McFadden, "some people just don't like highways through the woods." McFadden obtained Froshin's agreement that if the TRO was denied and a rapid hearing set for the preliminary injunction, the government would perform no substantial construction work in the interim. This was, in short, an informal TRO. Of course, the plaintiffs happily agreed as well.

Henry Froshin called me three days later. "Henry," I said, "I suppose you've called to capitulate!" "Well, yes," said he, "I suppose I have."

The government would agree to halt all future construction of the Talladega Scenic Drive and perform an EIS on the entire seventy-four miles if the plaintiffs would agree to permit the work on the pending connector segment to proceed. Of course we would. We had no objection to improved access to the state park.

Two weeks after it was filed, the Talladega Scenic Drive lawsuit was dismissed by written settlement agreement between the parties, and a milestone in Alabama conservation history was passed. The effect of this victory began making itself felt immediately. With construction halted indefinitely, the money that had been allocated to the Odum Trail segment was assigned to another project, and there was no inkling as to when, if ever, funding for the scenic drive might be restored.

A sullen *Birmingham News* complained that the government "had capitulated to the environmentalists. . . . It is time," the editorial thundered, "for Alabamians who want to see the drive developed to speak up. . . . For the voice of the environmental purist has been the only voice being heard, and that in opposition to the scenic drive. [Do] The Alabama Conservancy and the Sierra Club and a few others speak for every Alabamian?"[5]

CHAPTER EIGHT

Son of RARE

So, NOW WHAT? We had stopped the scenic drive, true enough, but the fact was that we had no real tool for protecting the Odum Trail backcountry except through the EIS process for the highway, which was expected to take a couple of years. The Federal Highway Administration, as the lead agency for the project, was given responsibility for producing the environmental impact statement, with input from the Alabama Highway Department and the U.S. Forest Service, but only the most naive could expect these agencies to change their minds about developing the road. The lawsuit simply stiffened their resolve to complete it.

We had one glimmer of hope for protecting the Odum Trail. During the Great Depression the Civilian Conservation Corps (CCC) built a dirt road down through the Talladega Mountains, known as the Skyway Motorway. It was over much of this old roadbed, in fact, that the Talladega Scenic Drive was to have been routed. However, in the vicinity of Cheaha State Park, the dirt road dropped off of the ridges down to the western base of Cheaha Mountain, and completely skirted the Odum Trail area before rising again to the ridge several miles to the south at a place known as Adams Gap (see map p. 50).

If we could convince the highway planners to follow this stretch of the Skyway Motorway, then there would be an alternative to destroying the Odum Scout Trail, an alternative that NEPA clearly required the planners to consider. The idea found no favor with scenic drive backers. "Everybody's well aware of it [the alternate route]," said retired colonel Claude Smith, spokesman for the TallaCoosa Highland Lakes Association. "It sounds like the same one I've heard about before, and the highway department and the forest service didn't think it was an acceptable route because it runs you way down the mountain."[1]

Meanwhile, in the northwestern part of the state, the U.S. Forest Service propounded a new threat to remaining wild areas in the Bankhead National Forest. In November 1976 it released a new ten-year timber management plan for the Bankhead that allocated virtually the entire 179,000-acre forest—excluding only the existing Sipsey Wilderness and a few developed recreation and administrative sites—to commercial timber production. In a gesture of profound insensitivity, it also provided for logging operations within the study area that Congress mandated when it targeted the West Fork Sipsey River for potential national wild and scenic river protection. The Bankhead timber plan set off a storm of controversy in the environmental community, with rumblings of another lawsuit in the offing.

What was desperately needed, both in the Bankhead and the Talladega Forests, was some instrument of reform, some mechanism through which a significant increase in Alabama national forest protected areas could be achieved.

That mechanism came from the administration of Pres. Jimmy Carter. Shortly after taking office in 1977, he ordered the Agriculture Department to revive the RARE process. RARE means "roadless area review and evaluation." Its purpose was to identify and recommend to Congress remaining wild areas in the national forests that might qualify for inclusion in the National Wilderness Preservation System. There had been an earlier RARE, undertaken in the 1960s at the direction of Congress when it passed the Wilderness Act, but the first RARE gave short shrift to potential wilderness in eastern national forests, on the theory that smaller acreages and historic evidence of man's entry into these roadless areas rendered them "impure" and therefore unqualified for permanent protection. As Forest Supervisor Dick Woody said in responding to criticism that his agency was antiwilderness, "A lot of people in Alabama might think that. We didn't think the Sipsey qualified for Wilderness designation under the law. The way it turned out, I guess it did."[2]

Carter's new initiative would compel the Forest Service to identify, and even recommend of its own accord, new eastern wilderness candidates. The process became known as RARE II, or as one wag put it, "medium RARE, the son of RARE."

Now completely entangled in national forest issues, I found the lure of coming wilderness campaigns impossible to ignore. When Mary Burks announced her plans to retire as executive director of the Alabama Conservancy, and its board of directors offered the job to me, I accepted with hardly a thought for the law practice I would abandon.

Natural area preservation in general, and RARE II in particular, was to be the dominant direction of the Conservancy. "The concept of this effort," said our press release, "is founded in our concern that the rapid development and commercial exploitation of our resources will soon leave the people of Alabama with little opportunity to enjoy the state's magnificent natural heritage. We must move to protect our natural areas, for they deeply influence the quality of our lives here."[3]

The Wilderness Society began a program to assist local groups in responding to RARE II. In the summer of 1978 the society's southeastern representative, Randy Snodgrass, joined us in Birmingham to help organize what was to become the Alabama Wilderness Coalition. The groups that joined the coalition were remarkable in their diversity, ranging from bowhunters to wildflower enthusiasts to traditional environmental organizations. Leadership of the coalition was vested in a steering committee, whose members were chosen through the all-American method of buying their way in. By providing funding, the Alabama Conservancy, the Sierra Club, the Birmingham Audubon Society, the local chapter of the National Speleological Society, and the Cahaba Girl Scout Council became the coalition's first steering committee. With only a few later changes in membership and shifts in financial responsibility, this group was to carry the burden of leading the RARE II wilderness campaign all the way to its conclusion in 1988.

Participation by scouting interests was vital to our hopes for protecting the Odum Scout Trail. The Boy Scouts of America proved to be a disappointment. A letter from an associate director of the Scouts' national office in New Jersey said,

> Thank you for your letter inviting us to join the Alabama Wilderness Coalition. While I can personally empathize with your efforts to preserve . . . the Odum Scout Trail, it has long been a policy of the Boy Scouts of America not to become involved on a national level with efforts that have political overtones. Our mission is to develop character, provide citizenship training and to promote physical and mental fitness among the youth of our country. We feel that this can best be accomplished without becoming identified with specific issues. . . . [W]e must respectfully decline your invitation to participate.[4]

Fortunately, the women of the Cahaba Girl Scout Council had a different concept of character and citizenship training, as did a number of individual Boy Scout leaders around the state, notably Mac J. Smith of Troop 15 in Montgomery. Through these sources, we were able to obtain statewide mailing lists of leaders and scouts for the several alerts that the coalition began to generate as the Cheaha wilderness campaign heated up.

Officially, RARE II in Alabama began with a call from the Forest Service for the public to nominate areas for consideration, together with release of the agency's own "inventory" of roadless lands. Surprising no one, the Forest Service found only about 18,500 acres in the state that it was willing to inventory of its own accord. Supplemented by Wilderness Coalition and other public nominations, the final Alabama inventory increased to "net" acres ("net" meaning public land only) of around 62,000—small enough at less than 10 percent of the state's national forest lands.

Each of Alabama's national forest units contained at least one inventoried area, but the Wilderness Coalition decided to direct all its efforts to two major candidates: A 30,000-acre expansion of the Sipsey Wilderness Area in the Bankhead National Forest, and, of course, a wilderness for the Odum Scout Trail, "smack dab in the middle," as we told the press, "of the proposed route for the Talladega Scenic Drive."

Our "nomination" included a hoped-for batch of national forest land totaling nearly 12,000 acres, encompassing all the public forest surrounding both sides of the Talladega Mountains down to Adams Gap and a spur ridge known as Cedar Mountain, along which the Odum Trail ran to its southern terminus at Pyriton. The Forest Service's inventory pared this down to about 8,400 net acres, but the major landscape features of mountain ridgelines remained intact.

The Forest Service had its own ideas of what to do about RARE II. Strangely, it began to generate a lot of heat, not in the vicinity of our priority areas, but in the Oakmulgee Division of the Talladega Forest, in central Alabama.

First, in a move that was more indicative of the agency's bias than it was effective, the Forest Service included in its own inventory an area in the Oakmulgee Division known as Perry Mountain. The mountain's only distinction was that it was regularly used by off-road vehicle enthusiasts. The attempt, apparently, was to stir up ORV opposition to new wilderness designations. It succeeded only in stirring up opposition to Perry Mountain as a wilderness candidate, which the Wilderness Coalition handled by simply joining.

Next was an administrative butchering of a RARE II candidate put forth on behalf of the Wilderness Coalition by Douglas J. Phillips of the University of Alabama, who lived near the Oakmulgee. Doug made an effort to create a coherent RARE II proposal along Big Sandy Creek, only to have the Forest Service reduce its acreage and then inventory it along with two additional small areas, all separated from each other, none constituting realistic wilderness candidates.

Then, the district ranger, Ben Fenton, met with a reporter for the *Centreville Press* and sounded the voice of doom. If the three areas near Big Sandy Creek, which together totaled about 6,800 net acres, achieved wilderness designation "a computer . . . projected losses for Bibb, Hale and Tuscaloosa Counties . . . at 63 lost jobs, a population decrease of 186 people, $658,000 lost in total annual income and a land value decrease of $1,123,000."[5] One wonders just what those 186 people were doing along Big Sandy Creek that, if told they could no longer do it, would force them to abandon it for better ground.

Just as alarming was an apparent conspiracy the *Centreville Press* reporter uncovered.

A look at the map hints that there is an intention to increase the wilderness . . . after it is established. Situated like the points of a strategic triangle or three beachheads on an island, the three proposed wilderness areas appear destined to outline the boundary of a single area . . . at least doubling its collective size at some future date. . . . Strong and effective are the various special interest groups pushing for wilderness. . . . The special interest pressure groups will push the wilderness designations through if they encounter no resistance.[6]

Ranger Fenton also made an effort to provoke antiwilderness sentiment among hunters. "Fenton said that if an area is declared wilderness . . . deer, grouse, quail, dove, turkey, rabbit and many songbirds would decline," leaving, in the words of the *Centreville Press*'s reporter, "nothing but squirrels." Well, maybe bears. But as Fenton pointed out, "since there were no bear [in the Oakmulgee], this would not be a factor."[7]

However, these shenanigans in the Oakmulgee Division occupied the realm of the irrelevant, since there was no realistic opportunity for a supportable wilderness proposal there. More important to the Forest Service was that it protect its pet project, the Talladega Scenic Drive.

During a public workshop designed to "rate" the wilderness characteristics of the individual roadless areas, rangers from the Talladega District Office began to press the merits of two new candidates north of Cheaha State

Park—1,900 acres in an area known as Blue Mountain, and 4,900 acres atop Alabama's second highest peak, Dugger Mountain. The message was clear: forget about the Odum Trail and the Forest Service would recommend these areas for wilderness designation.

Knowing, however, that we would be unlikely to accept such a trade-off, the Forest Service had another strategy. When the final state inventory was announced, the Odum Trail backcountry area had been split into two distinct RARE II units, separated exactly along the mountain crest by the planned route of the Talladega Scenic Drive. "There is a precedent for a highway dividing two wilderness areas," said George Gibbs, chief planning officer for the national forests in Alabama, ignoring the fact that no such highway existed in the area.[8]

In October 1977 the Alabama Road Builders Association convened a meeting with local officials and state legislators to plot a strategy "to combat environmentalists who are blocking construction" of the scenic drive. "I can't understand," said the owner of a company that had already built part of the highway, "why the businesses in the surrounding area don't push for this thing, because it means dollars, dollars, dollars."[9]

The U.S. Forest Service was happy to fill the void. What followed was a public antiwilderness campaign led by Forest Service land management planner George Gibbs, Assistant Forest Supervisor Jim Bylsma and Talladega district ranger Murray Johnson, which had the unintended and ironic effect of simply driving most impartial interests into the wilderness camp.

CHAPTER NINE

Are You Lying or Ignorant?

SOMETHING WAS DEFINITELY WRONG. On the telephone was U.S. Rep. Bill Dickinson, whose congressional district included southeast Alabama and the Conecuh (pronounced ka-NECK-ah) National Forest. There, the Forest Service had inventoried a little 2,700-acre RARE II area, almost by accident. It came into being during an early RARE II workshop, when one of our Wilderness Coalition leaders wandered over to the Conecuh Forest table and asked the rangers there to help him find an area to nominate. No one expected the nomination to be taken seriously.

Congressman Dickinson, however, was going to hold a public hearing. "We have some property owners whose land will be involved in the taking," he said, "and they're mighty unhappy about it. Besides, I seriously doubt any National Forest land in Alabama qualifies for Wilderness designation."

Dickinson had apparently also called the *Birmingham News,* as it immediately weighed in with a passionate editorial:

> Covington County residents are angry over what some see as a sneak attack by the federal government on the status of the Conecuh National Forest. They are angry at a plan to create a wilderness area in the forest which would force some landowners who lived on their land all their lives from their property. . . .
>
> The sudden craze for wilderness areas among a small group of the population has been difficult to deal with. The drive to set aside national forest lands for those areas has come mostly from groups of affluent people who "have it made" in a financial and cultural sense. . . .
>
> The idea of some wilderness areas set aside for fervent backpackers and addicts of nature-in-the-raw is fine, but where does it end?
>
> With millions of acres in Alaska being restricted from all exploration and development, it is time to halt the rush to create more wilderness areas until some kind of

cost benefit ratio can be established and decisions can be made on a rational basis rather than on the basis of sentiment.[1]

Then Randy Snodgrass of the Wilderness Society's office in Atlanta called to warn me that in parts of the South, the Forest Service was sending letters to property owners adjoining RARE II areas, giving the impression that their properties might be condemned if the wilderness proposals were enacted. Apparently, this had happened in the Conecuh, where Congressman Dickinson was playing it for all it was worth. We soon discovered that Talladega district ranger Murray Johnson had been doing the same thing.

The condemnation scare, as always seems to be the case with such things, arose out of technicalities in the law, but those technicalities also provided a solution and thus require explanation.

Being remote and undeveloped, Alabama's serious RARE II candidates were composed only of federally owned, national forest land, as the Wilderness Act specifically requires.[2] They were encircled by primarily dirt roads. Within the encircling roads were occasional tracts of privately owned land, called "inholdings," which backed up to national forest land, or, as lawyers sometimes say, intervened between the roads and the publicly owned land. In other words, the private land was on the periphery.

When the Forest Service compiled its RARE II inventory in Alabama, "for convenience" (so it said), it used the encircling roads as the RARE II boundaries, rather than only the public land lines, giving private property owners the impression that their lands were going to be within the proposed wilderness areas. The Forest Service obfuscated further by including those private inholdings in the tally of acreage it quoted for the RARE II areas, although it hedged on this a bit in its official publications by setting out both the "gross" acreage, including private inholdings, and the "net" acreage of only publicly owned land.[3]

Nothing in the Wilderness Act gives the federal government the right to condemn private property to "build" a wilderness. However, when it passed the Eastern Wilderness Areas Act in 1975, which included the original Sipsey Wilderness Area, Congress did grant the government the right to condemn private land "within" an eastern wilderness, on the condition that the land was being used by the owner in a manner inconsistent with the purposes of the Wilderness Act. The original Sipsey Wilderness does, in fact, have a few private inholdings that are completely surrounded by national forest land. In theory, these lands could be condemned, but in fact, they have not been.

In any event, a congressional sponsor of wilderness legislation (even Bill Dickinson, had he been so inclined) has the choice of completely eliminating the condemnation issue by designating the area under the original 1964 Wilderness Act, which grants no condemnation power at all, rather than under the 1975 Eastern Wilderness Areas Act. This was, in fact, what was ultimately done with respect to the Cheaha, Dugger Mountain, and Sipsey Expansion legislation. Certainly no one in the Wilderness Coalition, or any of the eventual congressional sponsors, had any desire to see anyone's land condemned.

Nevertheless, the Forest Service was playing off these obscurities to create as much disruption in the RARE II process as possible. Talladega district ranger Murray Johnson's letter to some fifty private landowners on the periphery of the Odum Trail RARE II area begins with the statement, "Your land may be directly affected," and ends with, "Because your land is involved, a special effort is being made to make you aware of this activity [RARE II]." The body of the letter itself, and the "fact sheet" accompanying it, contains a mish-mash of threats and hedges confusing enough to mislead anyone.[4]

Threat: "your land is involved . . . "

Hedge: "the boundaries of these *potential* wilderness areas are not fixed 'in concrete.' We anticipate some minor adjustments, i.e.—extensive private lands along the perimeter or exterior portion of the inventoried area should be separated."

Threat: "*What happens to private land that is included within the boundaries of, or is adjacent to a declared wilderness?*

"The 1975 Wilderness Act, Public Law 93–622, is one indicator of congressional land ownership policy within wilderness. . . . [U]ses of private land within designated wilderness are evaluated by the Forest Service for compatibility with wilderness management objectives. If found to be incompatible with wilderness use and the owners fail to discontinue such incompatible use, the Forest Service may acquire the private lands involved. In the case of residential or agricultural lands, the owners may elect to reserve occupancy and use for 25 years or for the remainder of their lives."

Hedge: "Wilderness designation in itself imposes no restrictions on the use of the private land within or adjacent to the wilderness."

Threat: "Even though wilderness designation applies only to Federal lands, any private land included within a roadless area designated by Congress as wilderness would be evaluated on the basis of existing land use. Owners are given an opportunity to discontinue incompatible uses."

Hedge: "Since the passage of this legislation in January of 1975, there has been no use of condemnation to acquire lands within wilderness areas by the Forest Service, nor is any contemplated at this time."

Threat: "However, such authority does exist and could be used under circumstances prescribed by law."

The brouhaha this caused among landowners on the eastern side of the proposed Odum Trail wilderness sent the *Anniston Star* in to investigate. "[District Ranger] Johnson says, 'I know full well' that in the end the private lands included for the temporary 'convenience' of drawing the [boundary] line down the road will be excluded from the RARE II area.... [H]e had already received orders to redraw the boundary to follow forest service lines. ... 'But still', said Johnson, 'we got people interested ... at least they know that the national forest is there and something is going on.'"[5]

Several of those "interested people" decided to write and give me "what for":

Dear Sir: I live in the Adams Gap area on land that was homesteaded by my ancestors 5 generations ago. This land is included in the proposed Wilderness Area.... Now you want to reach out and take our homes and land for your weekend play ground. How would you like for us to try to force you from your homes so we could come to town and live it up in them occasionally? ... [S]uch a place as this would be a haven for lawbreaks [*sic*] and idle persons who like to hide their deeds. It is my fervent hope that this proposal is voted down, but if it goes through we, the people, of this area will stand united as ever to protect that which has been ours through wars, depressions and all sorts of hard times through out our generations.

Dear Sir: We live in the Adams Gap ... area on land that was homesteaded by our ancestors four generations ago.... All these four generations we have worked to preserve its beauty. My Dad always taught us not to burn the woods and let the timber grow and be productive. Every body has to be productive, otherwise they are free loaders. This would be a heaven for the rapest to put his victim body and she could never be found, just to be eaten by the animals. Making a place for a group of people for so call pleasure, for a week-end is one thing, but when other hard working, law abiding people are distroyed at the same time is a horse of a different color.

Dear Sir: There are families who own private land in this area. They tell us that we will be able to stay here for 25 years or even our lifetime (this is not guaranteed) but we will not be able to use our land as we see fit. We cannot change the use from what it is today. We will not be able to cut our timber or even firewood off our own land.

Dear Sir: Come out here sometime and I will show you our country. I think you might agree that it is wild enough now.

It was inconceivable to me that the Forest Service could be so callous as to frighten people like this, but all I could do about it was to write them all back and tell them that their lands would never be included in the proposed wilderness—for all the good that did:

Dear Sir: I wonder if you are lying or ignorant.

CHAPTER TEN

Snail Darter for the Scenic Drive

IN EARLY JANUARY 1979 THE Forest Service released its final RARE II environmental statement. It recommended only a 7,000-net-acre expansion of the Sipsey Wilderness. The remaining 23,000 acres in the Bankhead, and the Odum Trail backcountry area, 8,400 net acres, were "allocated to further planning," meaning that their fate, as far as the Forest Service was concerned, would be decided in connection with the new Land and Resource Management Plan (the forest plan, for short) that was being developed for Alabama's national forests. George Gibbs was telling the press that he expected the plan to be completed by 1983.[1]

The time had come for the Wilderness Coalition to make a hard decision about its position on the Talladega Scenic Drive. The problem was demonstrated in a public campaign of letters to newspapers and elected officials by the TallaCoosa Highland Lakes Association, which read, in part: "Alabama may lose out on the development of its 'Smokies,' the 200,000-acre Talladega National Forest, due to environmentalists who want a 12,000-acre wilderness area established. The place in the forest that they have selected for this wilderness area is astride the right-of-way for the Talladega Scenic Drive.... It will be virtually impossible to go around a 12,000 acre wilderness area."[2]

The idea of paving the top of the Talladega Mountain chain was still as abhorrent to us as when we filed the scenic drive lawsuit, but the situation now was radically different. RARE II had given us the opportunity to gain permanent protection for the Odum Scout Trail backcountry. To achieve this required congressional action, which meant that we had to have the support of the U.S. representative for the area, Bill Nichols, and at least one of Alabama's senators. The Talladega Scenic Drive was Bill Nichols's personal project, and we knew from preliminary inquiries that he was displeased that we had shut his highway down. It was imperative that we find a way for Nichols to support a wilderness area.

We decided to take the position that if the highway south of Cheaha Mountain could be routed along the old Skyway Motorway in the valley until it arrived again at the crest at Adams Gap, and if Congressman Nichols would sponsor wilderness legislation for the Odum Scout Trail, the member groups of the Alabama Wilderness Coalition would all drop their opposition to the Talladega Scenic Drive. Granted, this would still leave the Hollins and Choccolocco Wildlife Management Areas vulnerable, but this highway battle had been going on for a couple of years now, and no one had heard a peep from hunting interests. The Wilderness Coalition had to go after what it could get.

The strategy began to show its political worth almost immediately. A group of us met with U.S. Senator Jim Allen to explain our RARE II goals. It was clear from his demeanor that he had been getting an earful from scenic drive supporters, and he was quick to tell us that he supported the highway. He listened politely to our appeal to protect the Odum Trail backcountry, but began to show real interest when we described the alternate route afforded by the old Skyway Motorway. "Well," he said, "I certainly don't have any loyalty to any particular route." Tragically, Jim Allen died of a heart attack shortly thereafter, but the attractiveness of our compromise had been affirmed. Our job was to keep hammering the point home.

The press was giving us a lot of opportunity to hammer, as the concept of a highway versus wilderness, particularly one containing a popular Boy Scout trail, was too attractive to ignore. "Highway May Send Scout Trail Down Road to Ruin," read the headline in the *Birmingham Post-Herald*. I loved it.[3]

The *Post-Herald*'s reporter, John Northrop, found an eloquent spokesman for the scouts in Robert Weaver of Talladega, who, shortly after the Scout trail was constructed, took on the job of promoting it in scouting circles by awarding Odum Trail patches to boys who completed the ten-mile hike. Weaver said,

> Based on the orders for patches, I'd say more than 30,000 boys have hiked the trail since it opened. They've come from all over the Southeast—Georgia, Florida, Mississippi, Louisiana. The appeal is that for 10 miles you don't see any roads, any signs of civilization. That trail is no cakewalk, and the only person who can get you to the other end is yourself. That's the way it used to be in the frontier days—hard work and personal effort. It's a fantastic experience for boys and I'd hate to see it change. I've received literally hundreds of letters from scout leaders telling me how much the accomplishment of hiking that trail has meant to their boys. If we had some other place

in this state to duplicate that kind of trail, this highway proposal wouldn't worry me so much. But there's no other place like it in the state.

The reporter then questioned the Forest Service about avoiding the Odum Trail by means of the Wilderness Coalition's proposed compromise route along the Skyway Motorway. "That idea was studied and rejected in the late 1960s," said planner George Gibbs. "One reason is that it would be unattractive when seen from a distance."

"Anybody who thinks a highway is pretty has something wrong with him," retorted a Boy Scout.

The Forest Service then unveiled its latest gambit: it moved the Odum Scout Trail.

In conjunction with its development of the Talladega Scenic Drive, the Forest Service planned to construct a new hiking trail, called the Pinhoti, to run the length of the Talladega Forest, roughly paralleling the scenic drive for its entire seventy-four miles. In preparation for highway construction south of Cheaha State Park, the Forest Service closed the traditional entrance of the Odum Trail at the state park, and rerouted the first few miles, moving it down on the mountainside to the west. From there, the new trail was to continue southwesterly on the side of the mountain, paralleling the scenic drive on top, until the two crossed in the vicinity of Little Caney Head peak. Originally, the Forest Service intended for the rerouted trail to be named the Pinhoti, but the highway controversy led them to call it the Odum Scout Trail instead, enabling them to assert that the trail would not be obliterated by the scenic drive.

"I would not let that trail be destroyed for anything in the world," Assistant Forest Supervisor Jim Bylsma told the press, emphasizing, nevertheless, how important it was that the scenic drive be built. The wilderness campaign reflected "a selfish attitude of some people who don't want to share the experience with others," he said. "We're not all young and vigorous and able to throw a pack on our backs and hike off into the wilderness."[4]

The trail relocation and attendant publicity resulted in a number of letters in protest to Congressman Bill Nichols, who replied by means of a form letter that read,

I appreciate you writing me in reference to the Talladega Scenic Drive and of your concern that the Odum Scout Trail be preserved.

I am familiar with the Odum Trail and find that it generally would skirt the proposed Talladega Scenic Drive along the top of Talladega Mountain for some three miles and would cross the Drive at that point and proceed south towards Pyriton for the remaining distance.

The Forest Service has assured me that the crossing point would occur on a flat ridge with ample vision in both directions as a safety feature for those who utilize the Trail.

It is my judgment that the Forest Service working with the many Alabamians who are interested in enjoying the natural beauty of this forest will be able to work together toward a satisfactory conclusion.[5]

"Congressman Bill Nichols appears to be practicing some sleight-of-hand deception on the Boy Scouts," read the *Anniston Star* editorial:

Nichols' letter is completely misleading. What has happened is that the Forest Service has moved to block off a portion of the real Odum Trail and has constructed an alternate trail, which it calls "The Odum Scout Trail." It is this trail Nichols refers to in his letter, not the real Odum Trail, which he apparently hopes everybody will forget. The [Talladega Scenic Drive] would, in fact displace large stretches of the real Odum Trail. Nichols and others supporting the [highway] have a right to their opinions, but the congressman is doing a disservice in the situation by distorting the facts.[6]

A few months later, Alabama attorney general Charles Graddick, together with Boy Scout Troop No. 15 and Explorer Scout Post 15, both led by Mac J. Smith of Montgomery, filed suit against the Forest Service in federal court in Birmingham, seeking to have the relocation of the Odum Trail set aside. The lawsuit was the work of Assistant Attorneys General Benjamin Cohen and Mark Brandon, and, representing the Scouts, Montgomery attorney James R. "Jim" Cooper Jr., a member of the Alabama Conservancy's board of directors. It alleged that the Forest Service had relocated the trail so that it could contend in its planning documents that the Talladega Scenic Drive would not destroy the Odum Scout Trail, thereby relieving it of the burden of considering alternate routes for the highway, as required by the National Environmental Policy Act.

Jim Bylsma was beside himself. The attorney general's office, he told the press, "[is just] making controversy. You can move a 5-foot wide trail easier than you can move a 40-foot wide highway. We [at the Forest Service] are beset from all sides by special interest groups. We have to try and satisfy everybody. The Conservancy is only interested in wilderness. The Boy Scouts are only a small percentage of the hiking public."[7]

On another front, the Forest Service had been catching hell all over the South because of the letters hinting at condemnation that it had sent private landowners on the perimeter of RARE II areas. In March 1979 Deputy Regional Forester Jim Sabin appeared in my office with a contrite Dick Woody in tow. To set the record straight, would I be willing to cosign with the Forest Service a new letter to those property owners on the eastern side of the Odum Trail backcountry who had earlier been provoked by District Ranger Murray Johnson? Of course I would, and on April 20, 1979, Dick Woody and I signed a "Dear Landowner" letter that read:

It was not the intent of the Forest Service to alarm you [with Johnson's letter]; however, it is clear from the number of protests received by both the Alabama Conservancy and the Forest Service that that did happen. This letter is an attempt to rectify that misunderstanding. Should the [Odum Trail] area be designated a wilderness by Congress, it will not include private land on the periphery of the area. The National Wilderness Preservation Act does not apply to private land. The Forest Service currently has no plans to acquire your land for Wilderness purposes. . . . We sincerely hope this letter will relieve any fears you may have had that your land was in jeopardy because of the National Wilderness Preservation program.

Sabin and Woody also assured me during our March meeting that the Odum Trail backcountry would be considered in all Forest Service planning as a single RARE II unit, rather than two areas separated by the scenic drive. However, within a few months, George Gibbs, the man whose responsibility it would be to develop the Forest Service's plans, was back in public touting his notion of "a highway dividing two wilderness areas."[8]

Every public dispute has its strange interludes, and now it was the turn of the Talladega Scenic Drive, with the entry into the fray of an organization called the Alabama Environmental Quality Association. The AEQA, as it was known, had been founded in 1968 by the Alabama Farm Bureau Federation as an outgrowth of an antilitter campaign. The group referred to itself, cryptically, as "Alabama's community betterment, public information and environmental agency for resource." For several years, the AEQA had been receiving state funding through the largesse of Gov. George Wallace, who described the organization as "the story of people in places like Opelika, Enterprise, Salem and Troy who are working together in a grand scale war against abandoned automobiles."[9]

In May 1979 the AEQA hosted an invitation-only meeting at Cheaha State Park designed to rally support for the Talladega Scenic Drive. Attended by Highway Department and Forest Service personnel, chamber of commerce types, and a few public officials, the meeting attracted good press coverage, supplemented by a singular AEQA press release entitled "RARE II, Possible Snail Darter for Scenic Drive."[10]

The snail darter reference was to the little fish whose protection under the federal Endangered Species Act had delayed construction of the controversial Tellico Dam in Tennessee, sending developmental interests and their political allies all over the country into a frenzy of demagoguery. (The previous year, for example, we had been treated to a state legislative race in which one candidate labeled the other a "snail darter extremist," a low blow indeed.) Now, the scenic drive supporters were invoking the specter of the snail darter because another federal endangered species, the red-cockaded woodpecker, was known to inhabit the Talladega National Forest. "We don't want to get into a situation where the redheaded woodpecker [sic] becomes a snail darter for the Talladega Scenic Drive," said state representative Gerald Dial.[11]

The Wilderness Coalition had never made an issue of the red-cockaded woodpecker or the Endangered Species Act, but the highway planners knew that the law required them to consider the road's impact on the bird in their environmental studies. In light of the then-current snail darter controversy in Tennessee, the scenic drive supporters hoped to stir up similar hysteria over the red-cockaded woodpecker. (They failed.)

Congressman Bill Nichols's in-state assistant, Bob Hand, attended the AEQA meeting, and he charged that the "Boy Scouts have been brought into the picture by the Conservancy to weaken the drive." However, he was quick to add that "Nichols is a 'very strong supporter of the Boy Scouts' and also is a supporter of wilderness and a supporter of the Scenic Drive."[12]

Jim Bylsma weighed in to complain of "'distortions and half truths' about the drive in newspapers around the state. 'There's been a letter-writing campaign also,' he said, 'and I for one don't like to see pressure put on public officials.'"[13]

Col. Claude Smith of the TallaCoosa Highland Lakes Association shared Bylsma's frustration, saying he "would appreciate if the news media would point out what it's costing to delay.... We may talk about building the Scenic Drive for $30 million, but if we keep dragging our feet and listening to a small group concerned only about a redheaded woodpecker [sic] falling out of its nest, it's going to cost $60 million."[14]

George Gibbs confirmed everyone's suspicion that "those who oppose the Scenic Drive have proposed RARE II as a way to block it," but he said they would not necessarily succeed, as "there is a precedent for a highway dividing two wilderness areas." Nevertheless, cautioned Gibbs, the scenic drive environmental impact statement and the Alabama forest plan would have to be completed before highway construction could resume. Gibbs now estimated that this planning would be finished "in two years."[15]

"Why take two years to do the whole thing?" fumed Claude Smith. "It's just woods. What's the use of a land management plan for a national forest? There's no use."[16]

Perhaps unintentionally, the AEQA invited an Odum Trail wilderness supporter, Margie McBride, operator of Camp Mac in the Talladega Forest. She "asked why people opposing the drive were not invited to the meeting. With all points of view represented, some kind of consensus could be worked out." State representative Gerald Dial replied, "The main thrust of the meeting is to keep before the public we want the drive. You don't invite the enemy in your camp when you're trying to win the battle."[17]

Since I was the most visible "enemy," the press called to ask me about all this. Describing the alternate route afforded by the Skyway Motorway, I said, "We have proposed an entirely reasonable compromise. If the developers can't accept a reasonable compromise, they deserve to have their project scrapped." Privately, I was delighted that the AEQA had gathered all these highway planners and supporters to publicly voice their prejudices. If all else failed, they had given us ample grounds to contest the impartiality of both the scenic drive EIS and the Alabama forest plan in court.[18]

There was nothing left now but for the executive director of the AEQA,

Martha McInnis, to give a coherent explanation of her group's support for the highway: "The Scenic Drive is very unique to the southeast. People are involved, not land condemnation. The Scenic Drive would take what we've already got in land management and use it."[19]

This meeting proved to be the only flirtation by the AEQA with the wilderness issue. Later calling itself EnviroSouth, it ceased to exist when its public funding was eliminated during the first administration of Gov. Fob James.

CHAPTER ELEVEN

Breakthrough

ON THE TELEPHONE FROM Washington, D.C., was Winston Lett, a young Lee County attorney who had recently been named Congressman Nichols's legislative assistant. He wanted to know what our true intentions were. "Everybody's telling us that you proposed this Wilderness just to block the Scenic Drive."

Not so, I said. "We'll drop our opposition to the highway if it can be rerouted to follow the old Skyway Motorway in the valley and if Mr. Nichols will sponsor wilderness legislation for the Odum Scout Trail area." Lett encouraged me to keep thinking along the lines of a compromise. "The boss wants to get the Scenic Drive built before he retires, and he doesn't want to see any more lawsuits. But," he cautioned, "he thinks that the highway ought to have a view."

Encouraged by this peace feeler and by growing support in Nichols's district—including key editorials from the *Anniston Star* ("a priceless gift for future generations") and the *Talladega Daily Home* ("we can have the drive and wilderness, too")—I felt it was time for a personal meeting with the Congressman. Nichols agreed to see me at his home in Sylacauga in early July, during Congress's summer recess.[1]

George Gibbs had been telling the press that the Forest Service and Highway Department were developing up to four "alternatives" for the scenic drive in their planning process, so I wrote to his boss, Dick Woody, telling him of my upcoming meeting with Congressman Nichols and asking him to send me a map of these alternatives so we could discuss them in July. Woody replied, "Though I'd like to provide the map you requested, I don't have one and I sincerely believe it is premature for any of us to get too attached to any particular alternative. To try to negotiate an alternative would preempt an open, objective decision based on the best interests of all publics. The NEPA process for both the Scenic Drive and our Land Management Plan will incorporate appropriate analyses from which the most feasible alternative can be determined." Clearly, the Forest Service had no intention of surrendering control of the project, even to a U.S. Congressman.[2]

On July 2, 1979, my thirty-fifth birthday, I pulled a chair under Bill Nichols's breakfast room table and hoped that I could convince him to resolve the Odum Trail controversy. It was obvious that he was unhappy with what the Wilderness Coalition had done to his highway project, but he was quick to tell me, "I'm a conservationist too, you know."[3]

Nichols wanted straight answers. Were we trying to kill the scenic drive?

Originally, I admitted, that had seemed the only option when we filed the scenic drive lawsuit in 1976. Now, we saw RARE II and the alternate highway route afforded by the Skyway Motorway as the means for the coalition to drop its opposition to the road.

"Why," he asked, "couldn't the highway and the trail coexist on the mountain top? The Forest Service tells me that the Odum Trail is not a very good one and the new one has better views. We could make a Wilderness Area for it that runs from the highway down the side of the mountain."

The issue of the Odum Trail was not which path had the better views but the invasion of the mountain wilderness it traversed with a paved highway. The ridgeline was so narrow that running a highway down it would forever destroy a mountain wilderness experience that could not be duplicated anywhere else in Alabama. Besides, if tourism were the objective, having a National Wilderness Area adjacent to Cheaha State Park would add to the attraction.

"Now, if I agree to sponsor a wilderness for this area, are you going to be coming back to me again a year or two from now, asking for more wilderness in my District?"

"No, sir," I said, "this is it."

"Well," said Nichols, "here's another problem. If we follow the old CCC road, the highway might have to go through heavy use areas in the State Park, and that could be dangerous."

"Congressman," I said, "I'm no highway engineer, but I'll bet if the Highway Department spent half as much time figuring ways to by-pass the heavy use areas as they've spent opposing the alternate route, this wouldn't be a problem."

"Well," he replied, "I think the Scenic Drive has to have a view."

There it was again: a view. This was going to be our biggest obstacle with Nichols. "We understand that the Forest Service is working up some new alternatives that may be feasible," I said, "but unless someone will take the lead in getting a fair compromise worked out, this thing is probably headed back to the Courts."

"No, no, let's not let it come to that," said Nichols, shaking my hand. Congressman, I thought to myself as I left, you're the only one who can stop it.

It never occurred to me that a solution to the scenic drive dispute might come through the Alabama Department of Conservation and Natural Resources, especially its Game and Fish Division. These entities historically were linked with the timber industry, for several reasons. Managing the timber in certain ways improves habitat for popular game species such as turkey and deer, and several large corporate timber interests had entered into cooperative agreements with the Conservation Department to manage some of their lands for regulated public hunting as wildlife management areas. A great many of the people employed in the timber industry, both white and blue collar, were sportsmen who liked to hunt, as, of course, did many Game and Fish Division employees. It was a natural fit, but the mind-set of managed forests devoted to the propagation of big game species was too often indifferent, if not downright hostile, toward "nonconsumptive" values such as habitat for songbirds and wilderness preservation.

But in this summer of 1979, a new administration was installed in Montgomery, that of Gov. Fob James, who appointed one Richard A. "Dick" Forster as commissioner of the state Department of Conservation and Natural Resources. Douglas McGinty, a biology professor at Huntingdon College in Montgomery, and a member of the Alabama Conservancy's board of directors, asked Forster if he would come to Doug's home for beer and snacks with leaders of the Alabama Wilderness Coalition to discuss the scenic drive.

Forster was accompanied at our meeting by his department's most powerful division head, Charles Kelley, director of game and fish, whom many regarded as commissioner in fact, if not in name. Kelley was straightforward in explaining why he wanted to meet with us. He was aware of a nationwide decline in "consumptive" forest users (hunters) and a corresponding increase in "nonconsumptive" users (nature lovers). Many states had increased their conservation-related revenues through such means as taxes on binoculars and camping equipment, income tax refund checkoffs, and the like. Kelley wanted to do the same in Alabama, and he wanted the coalition's support.[4]

Kelley also delighted us by saying that the Game and Fish Division had serious problems with the routing of the Talladega Scenic Drive through Choccolocco and Hollins Wildlife Management Areas. It would be to his agency's advantage if he could help produce a political settlement that would restrict or eliminate development in the two popular hunting areas. We were all for that. Although reserved at first, as he had been hearing the usual mantra of our RARE II proposal being a ruse to sabotage the scenic drive, Commissioner Forster warmed to our idea of an alternate route for the highway, and by the meeting's end he decided to offer his services as a mediator in hopes of concluding the dispute. He and Charles Kelley went to work.

Kelley took Congressman Nichols on a tour of the old Skyway Motorway and found Nichols receptive to compromise. Likewise, officials at Cheaha State Park and in the Alabama Highway Department were willing to talk. Only the U.S. Forest Service refused to cooperate. "We met with them for about two minutes," Dick Forster told me. "When we asked for their opinions about alternate routes, they said that they wouldn't discuss it until their Land Management Plan was finished, so we just rolled up our maps and walked out."

We decided to give the Conservation Department some favorable publicity. "The attitude of Commissioner Forster and Game and Fish Director Kelley toward reaching a solution to the Talladega Scenic Drive controversy has been like a breath of fresh air," I told the press. We were rewarded with a *Birmingham News* editorial commending Forster for his efforts: "Perhaps through his initiative, a compromise can be worked out that will allow this important roadway to be finished while dealing with the legitimate concerns of conservationists."[5]

"The legitimate concerns of conservationists"! This, from a newspaper that had been dismissing us as "environmental purists" and "addicts of nature-in-the-raw." Change was definitely in the air.

Before August was out, Forster was back with a specific proposal that he was sure Congressman Nichols would accept:

1. The Talladega Scenic Drive would be completely eliminated in the Choccolocco Wildlife Management Area, and certain restrictions placed on its development in the Hollins Area.

2. South of Cheaha State Park, the drive would follow, instead of the old CCC road in the valley, the 1,400-foot contour line on the side of the mountain, in order to give it the "view" that Bill Nichols required.
3. There would be a wilderness area created for the mountain top and the Odum Scout Trail.[6]

From the perspective of our 1976 scenic drive lawsuit, this proposal offered an overall solution to the highway dispute more successful than we could have imagined when we filed the litigation. The compromise route south of Cheaha would run some 600 to 950 feet below the ridgeline, providing a buffer for hikers on top. Most importantly, the plan would attract Bill Nichols's sponsorship of National Wilderness Area legislation for the Odum Scout Trail. The Alabama Wilderness Coalition quickly and unanimously endorsed the proposal.

The U.S. Forest Service did not. It continued to hold out for completion of the scenic drive EIS and the Alabama forest plan, which would focus upon "a number of proposed routes," according to George Gibbs. "Gibbs says the forest service wants to hear public comments on all the alternatives before making the decision on the scenic drive. He has been critical of the so-called 'compromise' route pushed by Nichols, Forster and the environmental groups. 'There are a lot of end runs being made trying to prejudge and predetermine what happens,' he complains."[7]

Then, quite abruptly, my personal leadership of the Cheaha Wilderness campaign came to an end, as the Alabama Conservancy simply ran out of money and was unable to support a full-time executive director, even at the modest salary I was drawing. I had to resign and do something about earning a living. For a year I returned to the private practice of law, then dove back into the wilderness fight through a retainer arrangement with the Birmingham Audubon Society, which was funded in part through a challenge grant from my father's estate. Since the Odum Trail fight seemed nearly finished, my new priority was to be the expansion of the Sipsey Wilderness in the Bankhead National Forest.

Meanwhile, the Conservancy's volunteer vice president for conservation, Jim Cooper, the Montgomery attorney who was representing the Scouts in the lawsuit filed by the attorney general's office to stop the relocation of the Odum Trail, felt it was important to get Congressman Nichols's signature on Dick Forster's plan. In February 1980 a memorandum of understanding, detailing the specifics of the compromise, was signed by Jim on behalf of the Alabama Conservancy, by Conservation Commissioner Dick Forster and, most importantly, by U.S. Representative Bill Nichols. The highway dispute at last seemed to be resolved.

Or was it? The U.S. Forest Service, as George Gibbs was to remind everyone in the months to come, was not a party to the memorandum of understanding.

CHAPTER TWELVE

Deja-Breakthrough All Over Again

IF THE MEMORANDUM OF UNDERSTANDING was flawed from the Forest Service's standpoint because they did not sign it, it was also flawed from our perspective because it did not specifically say that Bill Nichols would sponsor wilderness legislation, or when. As 1980 expired with no further activity, conservation leaders around the state became concerned that the initiative would return by default to the highway planners, who continued to quietly develop their scenic drive EIS and the Alabama national forest plan.

The Wilderness Coalition had lain dormant for more than a year following my resignation from the Alabama Conservancy. Now, in early 1981, I was back in operation as director of natural area preservation for the Birmingham Audubon Society, beginning to work actively on the Sipsey expansion. The Conservancy's newly elected vice president for conservation, Steve Spencer, called the member groups together to effect a rebirth of the Wilderness Coalition. Since I was already tasked and funded for the new Sipsey campaign, the Conservancy agreed to continue with the Odum Scout Trail effort under the leadership of one R. Michael Leonard.

A transplanted North Carolinian, Mike Leonard was a young associate attorney with Cabaniss and Johnston, an old line Birmingham business law firm. Over the next two decades, Mike would become one of the South's most effective, and most recognized, conservation leaders. He won the Sol Feinstone Award from the State University of New York and was the Alabama Conservancy's Conservationist of the Year for his wilderness work in Alabama. His specialty became negotiating deals and lobbying funding for land acquisition projects. Panthertown Valley near Cashiers, N.C., is now national forest land and a popular recreation destination because of Mike's skills and hard work. His dream, however, has long been to extend the Appalachian Trail (AT), in fact if not in name, from its current terminus at Springer Mountain,

Georgia, eastward into the mountains of North Alabama, then down the Talladegas by means of the Pinhoti Trail, to the true terminus of the Appalachian chain in Coosa County. Mike continues to succeed in getting the land acquired, and the trail constructed, bringing the reality of this connector to the AT closer every year.

The third organization to influence the wilderness effort through the 1980s was the Sierra Club, which now had its own Alabama chapter. However, Wally Retan had moved on to other interests and no comparable local leader had yet surfaced. Fortunately, the club had recently opened a southeastern regional office and had hired a native Alabamian, James M. Price of Huntsville, as its first southeast regional representative. Jim had served on the Alabama Conservancy's board of directors and was already contributing to the wilderness campaigns when hired by the Sierra Club.

Jim Price, Mike Leonard, and I became the triumvirate, if you will, that directed the Wilderness Coalition during the early 1980s.

Mike's new inquiries to Bill Nichols's office led the congressman to schedule a meeting in April 1981 at the Alabama national forest supervisor's office in Montgomery, with representatives of the Wilderness Coalition and the Alabama Highway Department present, to discuss the status of the scenic drive. Nothing had been done in eighteen months, said the Highway Department official, since both the scenic drive EIS and the national forest plan were still pending. Nevertheless, the department had recently received funding for engineering work and could begin when they knew for certain which route the highway was to take.

This led George Gibbs of the Forest Service to again make his pitch for waiting until the Alabama forest plan was complete; but his boss, Dick Woody, surprised everyone by saying that a wilderness bill was what was needed to "unhook" the time problems caused by planning delays. In the end, a consensus was reached that the western boundary of the new wilderness area would be the 1,400-foot contour line, or any other location farther west that the Highway Department engineers found to be feasible. Mike Leonard agreed to draft a wilderness bill to this effect for Congressman Nichols to review.

In August 1981 the Alabama Highway Department notified Bill Nichols that the 1,400-foot contour was unworkable from an engineering standpoint and recommended that the scenic drive be routed along the old Skyway Motorway in the valley, as the Wilderness Coalition had been advocating all along. "Winston Lett of Nichols's office said . . . he assumed what the highway department recommended would be followed. 'So for us that could be a minor problem to work out.'"[1] A draft wilderness bill was in the works.

Four months later, Mike Leonard received a surprise from the Forest Service, a letter from Dick Woody stating that he and George Gibbs had met on December 23 with Highway Department officials and Winston Lett of Bill Nichols's office:

> The highway department has studied the proposal to route the Scenic Drive on or about the 1400 foot contour and has determined that this alternative corridor is infeasible. . . . The conclusions reached by those at the December 23 meeting were: . . . We will eliminate the 1400 foot contour as an alternative corridor for the Scenic Drive. . . . We are not now in a position to support or oppose legislation to designate National Forest land as Wilderness. . . . Mike, since we are closing in on the due date for the Forest Plan and the state's EIS for the Scenic Drive, I feel we should adhere to [these] processes.

"Wilderness Bill Shelved," read the *Anniston Star*'s front-page headline:

> A bill intended to designate a portion of Talladega National Forest as a federally protected wilderness area has apparently been scrapped for the immediate future by U.S. Representative Bill Nichols of Sylacauga. Press Secretary Tom Eiland said Nichols will likely not introduce the . . . bill because of problems concerning the eventual location of the proposed Talladega Scenic Drive. . . . Nichols plans to wait until next year [1983], when the U.S. Forest Service intends to complete a land-use management plan for the Talladega National Forest, and then hold public hearings on the entire question of highway location and wilderness boundaries. . . . Eiland says some citizens are still interested in putting the highway on top of the ridge. "What Mr. Nichols says is that they certainly have a right to be heard and it should be discussed. . . . The wilderness area's been there for a long time, and I'm sure over the next 11 months it isn't going to go away. . . . What we need to do is just go back and kind of smooth out some rough edges and it'll come out . . . two, five years from now, or whatever, we'll hopefully have a wilderness area down there."[2]

I'd never seen Mike Leonard so mad. Nichols's office had given him no warning of the congressman's reversal of position. Substantially at his own expense, Mike developed an alert on the Odum Trail situation, tactfully laying the blame on bureaucrats rather than Bill Nichols, saying that "the Forest Service is trying to get Congressman Nichols to dishonor his agreement," which was true enough. Mike asked every group he could think of to circulate the alert. The Birmingham Boy Scout Council readily agreed, but Mike's real coup was talking Anniston's highly reluctant Choccolocco Council into distributing the appeal for letters. Why reluctant? The council's chief executive told me later that he felt a "loyalty" to the Forest Service "because they let us build the Odum Trail in the first place."[3]

Mike's appeal very effectively reopened the public controversy, but in the end, four other factors combined to tip the balance in favor of Odum Trail preservation.

The first was that old Dick Woody up and retired on us. Alabama was getting a new national forest supervisor, one Joe J. Brown, who was transferring from a California forest. Without the baggage with which years of scenic drive advocacy had burdened Dick Woody, Brown entered the situation with an open mind.

The second came courtesy of Ronald Reagan's new administration. In December 1981 it issued a directive prohibiting the Forest Service in the southern region from recommending *any* RARE II "further planning" areas for wilderness designation in the upcoming forest plans. Instead, the plans could only recommend that RARE II areas be designated by Congress as wilderness study areas. This meant there would have to be an act of Congress, signed by the president of the United States, directing the Forest Service to perform yet another study, to recommend whether or not there should be another act of Congress, signed by the president of the United States, designating a RARE II area as wilderness. Mike Leonard sent a copy of this directive to Congressman Nichols, saying that this nonsensical procedure "would pile plan on top of costly plan and put the final outcome of the Odum Scout Trail question four or five years on down the road."

A third factor was that Congressman Ronnie Flippo of Florence was ready to introduce legislation to expand the Sipsey Wilderness Area and he wanted Nichols to join him in sponsoring an Alabama Wilderness Act. "I don't think there's any doubt," Mike Leonard said later, "that if Flippo hadn't been pushing his Sipsey bill at the same time, Nichols might never have sponsored Cheaha legislation."

But perhaps the decisive factor was that Bill Nichols wanted to be perceived as a conservationist and a man of his word. As the *Anniston Star* editorialized, "We believe that Nichols should have stayed with his commitment. Since he didn't he is obligated to resolve the matter (again) as promptly as possible."[4]

So when Joe Brown took over the forest supervisor's job in early 1982, Congressman Nichols once again called all the parties together to a meeting at the Forest Service office in Montgomery. The outcome was surprisingly noncontroversial. Nichols came down hard on the Reagan administration's new policy for dealing with RARE II areas, saying he wanted to avoid all the bureaucratic delay. Joe Brown replied that the best way to do that would be through wilderness legislation and his office would be happy to help. Everyone agreed, without dissension, on the scenic drive route as it exists today: from the top of Cheaha Mountain, it would run south in a straight shot down the western slope, avoiding the heavily used lake and campground at the foot of the mountain, until it reached the old Skyway Motorway in the valley, following it until it rose again to the ridge at Adams Gap.

Then occurred one of the most memorable events of the long scenic drive dispute. Although Nichols had agreed with the Alabama Department of Conservation that the Talladega Scenic Drive would not be built in its northern reaches through the Choccolocco Wildlife Management Area, the Highway Department had continued to plan for the entire seventy-four-mile route and had appeared at this meeting with all its cost estimates in hand.

I have an indelible memory of Bill Nichols seated at the Forest Service's conference table with two Highway Department officials standing behind and leaning over to hand the congressmen their cost work-ups.

"What are these here?" asked Nichols, picking up a sheet. "Are these the costs for the route we've agreed on?"

"Oh, no sir, they're for the northern segment above I-20."

"Oh," said Nichols, taking a black felt marking pen and making a big X on the Highway Department's document. The two Highway Department men stood there, wide-eyed and gaping. Thus died the northern segment of the Talladega Scenic Drive, once and for all.

On May 21, 1982, Congressman Nichols introduced his legislation creating a 6,780-acre Cheaha Wilderness Area, completely encompassing the original Odum Scout Trail and the main ridgeline of Talladega Mountain down to Adams Gap. His bill was combined with Ronnie Flippo's Sipsey legislation to become the Alabama Wilderness Act of 1982. The act passed the House of Representatives on August 1, 1982, and was sent to the Senate for action.

There, it confronted a growing controversy over the failure of Alabama's two U.S. Senators to support the Sipsey Wilderness expansion. Howell Heflin, Democrat from Tuscumbia, took the position that it was out of his hands, as the Republicans now controlled the Senate, and it would be up to Republican senator Jeremiah Denton of Mobile to see that the legislation received hearings and was acted upon. In the end, Denton effectively killed the Sipsey legislation by sponsoring only the Cheaha proposal. Denton's staff, new and inexperienced, had difficulty handling the procedures involved in moving legislation through the Senate, so Bill Nichols asked Senator Heflin to step in to get the Cheaha Wilderness Act approved in the waning days of the 97th Congress. "It was like winning a football game with a field goal in the final seconds," said Nichols.[5] President Ronald Reagan made it the law of the land on January 3, 1983, and Alabama had two national wilderness areas.

Now it was time to celebrate. Bill Nichols was a wounded World War II veteran, and he made a congressional career out of championing the cause of the military from his seat on the House Armed Services Committee. He even wore clothing that resembled a uniform, khaki colors, shoulder straps on his shirts, and the like. His aide, Winston Lett, helped Mike Leonard plan a wilderness dedication at Cheaha State Park in August 1983 that resembled an army ceremonial, with the Fort McClellan Fourteenth Army Band playing John Phillip Sousa, and color guards made up of Boy and Girl Scout Troops from the local area. During the event, Mike unveiled a granite slab he had commissioned to honor Nichols. The slab was actually a tombstone, and the carver did not know how to spell "wilderness," so we ended up dedicating the Cheaha "Wildernes." No one seemed to mind; it was a happy occasion.

During the course of the scenic drive dispute, both the Forest Service and the Alabama Highway Department often asserted that the mountaintop route would be much cheaper than following the old Skyway Motorway in the valley, although both admitted that no one knew for certain. Once the agreement was reached on the scenic drive, however, the Highway Department let it be known that the ridgetop route down the Odum Trail would have cost $1.4 million dollars more than the valley route.

Congressman Nichols died in office in 1988. Before his death, he succeeded in procuring funding only for the segment of the scenic drive that runs in the valley west of the Cheaha Wilderness Area to Adams Gap, where the pavement now ends. Further construction of the road—which is today called the Talladega Scenic Byway—has never since been a priority of the Alabama congressional delegation.

Jim Bylsma and George Gibbs both retired from the U.S. Forest Service. Although Bylsma's highway proposal failed, his dream of a Pinhoti Trail through the Talladega National Forest is a reality, and, thanks to Mike Leonard, may someday be connected with the Appalachian Trail in Georgia. The Odum Trail still exists, running first as a segment and then a southeasterly spur of the Pinhoti.

The attorney general's lawsuit against the Forest Service for relocating the Odum Trail was rendered moot by the Cheaha wilderness legislation and was ultimately dismissed without fanfare.

My own representation to Bill Nichols that we would not ask him for another wilderness area in his district proved to be a mistake. While I was sincere at the time, neither I nor any of the other Wilderness Coalition leaders had ever visited Dugger Mountain. We viewed it only in terms of the Forest Service's attempt to substitute it for the Odum Trail area. It never occurred to us that it might be a legitimate wilderness candidate in its own right. We were wrong—I was wrong. It's a mistake ever to write off a candidate for preservation once it's been identified by the Forest Service.

Mike Leonard returned his tombstone to the carver and had him add another *s* to "Wildernes." Now, what to do with it? Mike, who has never lacked chutzpah, convinced a group of his friends (Guy Arello, Maggie Mead, Kathy Stiles Freeland, Tony Cooley, Rebecca Falkenberry, Jim Redwine, Tom Roberts, my wife, Virginia, and me) to haul the damned thing—it weighed a couple of hundred pounds—on a sling strung between two poles that we carried on our shoulders, up the Odum Trail to the top of Hernandez Peak, where it resides today, overlooking the trail's descent toward McDill Point. We toasted the Wilderness with champagne and christened the granite slab the P Stone. I'll never tell why.

PART THREE

The Sipsey Wilderness Expansion

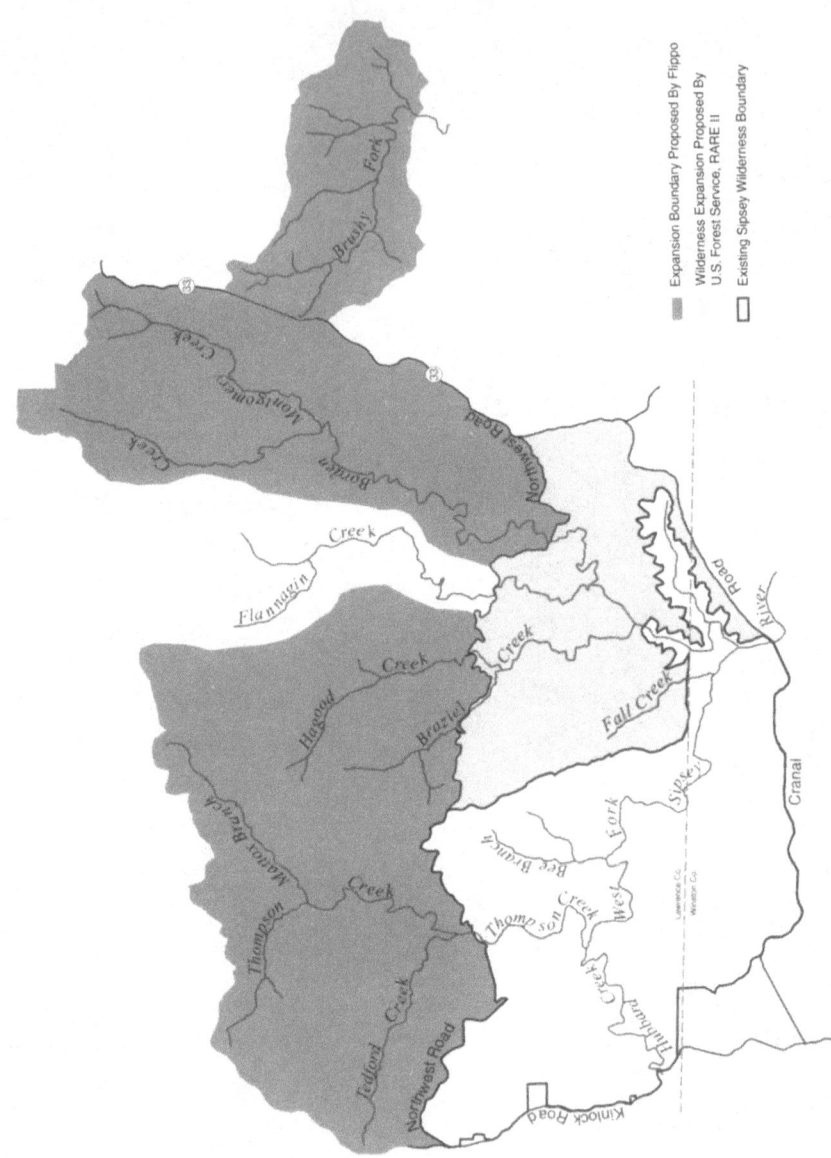

Sipsey Wilderness Expansion as Proposed by Rep. Ronnie G. Flippo, 1982

CHAPTER THIRTEEN

Perceptions

THE BATTLE TO EXPAND THE Sipsey Wilderness Area was fought, initially at least, in terms of its supposed negative impact on the local timber industry. Although there was overwhelming evidence to the contrary, this did not prevent the expansion's opponents from predicting doom if the wilderness were to be enlarged.

The strategy adopted by the opponents had three facets: exaggerate the economic impact of wilderness "reservations"; belittle the quality of the lands to be protected; and malign the people who sought their protection. The nasty little devil called condemnation also continued to raise its ugly head from time to time.

The drumbeat started in the late 1970s, when the Carter administration launched its RARE II program. Not surprisingly, the U.S. Forest Service was willing to fire the first shot. Bob McCallum of the forest supervisor's office in Montgomery told the press, "A private individual has the luxury of calling anything a wilderness, but the Forest Service has to go by what's written in the law. Some of the land [in Alabama's RARE II inventory] obviously doesn't meet the definition, but the guidelines established by Congress have been somewhat relaxed so that all of these areas could be included."[1]

It was true enough that all the land in Alabama's RARE II inventory did not qualify for wilderness designation, but it was not true that the standards established by Congress for statutory protection had been "relaxed" in any way. This was just another way of expressing the agency's discredited purity doctrine. As Alabama forest supervisor Dick Woody had said after the original enactment of the Sipsey Wilderness, "We didn't think the Sipsey qualified for Wilderness designation under the law. The way it turned out, I guess it did."[2] Woody got one thing right: Congress, not the Forest Service, defined what qualified as statutory wilderness.

The purity dispute was at least an honest difference of opinion about policy. The Alabama Farm Bureau Federation decided to refocus the debate:

A state Farm Bureau official said Monday that increasing the size of protected wilderness in the Bankhead National Forest would prove economically detrimental to the immediate area and its timber-related industries.

Furthermore, Dick Fifield maintained that federal programs such as the Roadless Area Review and Evaluation II . . . were threats to private property as the country knows it.

"There's a movement in this country and a feeling that you don't own your land but only a negotiable package of rights to it," Fifield said.

"It's become an issue of taking and taking too much. No one will argue the value of wilderness or scenic rivers, but there is a dispute over the size and scope of RARE II.

"How much is this country going to take out of economic production," he asked. " . . . If you have a resource, you need to utilize it to its highest point. It's like planting corn and never harvesting it."

Fifield said that the wilderness movement had originated with a group of people having "noble ideals" but "who would have us living in mud shacks and eating roots and berries."

"How many people are going to put a pack on their backs and go into the wilderness," he asked. "I don't think anyone has trod across the wilderness we have now."[3]

Winston County agent Robert Murphy chimed in with, "the wilderness areas are of little worth to the area. You get absolutely nothing from it. It just stands there with no economic value for anyone."[4]

Frank McAlpine was the owner of Bankhead Forest Industries, which owned and ran the sawmill town of Grayson near the Winston–Lawrence County line. He said, "The wilderness area already set aside isn't worth a nickel to anybody. . . . That wilderness is a briar patch. Nobody's going to go walking through there. . . . [I]f you like hiking, there are plenty of roads. You can walk yourself to death if you want."[5]

The more moderate Alabama Forestry Association—the primary state trade association for the timber industry—said it would oppose "limitless" wilderness proposals resulting from RARE II, and even this was an obvious exaggeration, as were all these opponents' claims. Alabama had a tiny amount of publicly owned land, the largest block of which lay in the some 641,400 acres of national forest. As originally proposed, the enlargement of the Sipsey Wilderness to 42,700 acres, plus the 6,700 acres contained in the Cheaha

Wilderness, amounted to only 8 percent of Alabama's national forests, and less than two-tenths of 1 percent of total commercial forestland in the state of Alabama. Indeed, all lands in the RARE II inventory, whether qualified for statutory designation or not, amounted to no more than three-tenths of 1 percent of the state's 21.3 million acres of forest.

Nevertheless, these early days were a time for bridge-building, and the Alabama Forestry Association met us more than halfway by sponsoring a dinner with RARE II leaders in December 1978. At its conclusion, the association's Bob Wiggins "predicted that even if his group doesn't endorse all proposed wilderness areas, some 'middle-ground' can be found."[6] Unfortunately, as the Sipsey expansion dispute heated up, moderate elements within the Alabama Forestry Association were silenced by its leadership, and it would be several years before the group made good on its promise to help find a middle ground.

As for the Alabama Farm Bureau, we confronted Dick Fifield about his claim that wilderness advocates wanted "us living in mud shacks and eating roots and berries," and his outrageous use of the condemnation scare, there being of course no "taking" or "threats to private property" involved. He apologized, somewhat sheepishly, saying that he was just participating in the Farm Bureau Federation's national campaign against RARE II. In any event, this proved to be the end of his group's public foray into the Alabama wilderness debate, although it continued to lobby privately against the Sipsey Wilderness expansion.

On the positive side, several studies were released during this early period that supported the need for additional wilderness designations, a couple of which came from the Forest Service itself. When it released its final RARE II EIS in January 1979, the Forest Service concluded that wilderness designations in Alabama would have "insignificant economic effects."[7]

The Forest Service's regional office in Atlanta reported that

[t]he public demand for wilderness has increased in recent years. The public has expressed a need for wilderness recreation opportunities as well as a wilderness system that enhances and preserves various kinds of land forms ecosystems, wildlife and fish populations. Wilderness visits in the [Forest Service's] Southern Region [which includes Alabama] increased an average of 5% from 1975 to 1979. In 1979, a 9% increase was recorded.... Based on an average 4% annual increase in recreation use of wilderness in [the southern region], more wilderness acreage will be needed in about 10 years.[8]

Locally, in January 1979 a group of North Alabama regional councils of government completed a cooperative study titled *Land Suitability Analysis of the Bankhead Forest and Smith Lake Area of Alabama*. Huntsville native Jim Price, who was employed at the time by the Northwest Alabama Council of Local Governments and later became the Sierra Club's first southeastern regional vice president, was instrumental in researching the report's conclusions:

> Given the demand for increased recreational opportunities in the [Bankhead–Smith Lake area], and the relative abundance of commercial forestland in the rest of Alabama, it appears that there may be a maldistribution of resources within the Bankhead Forest. . . . Consideration should be given to the creation of more wild and scenic areas, protected from commercial timber operations and mining activities. . . . Doubling or tripling the area reserved for public recreation of all kinds would have virtually no economic effect on the surrounding area, and would provide a much better balance of use. . . . The areas most appropriate for inclusion in any expansion of wild or scenic lands have already been identified [by] RARE II.[9]

So, if, indeed, the proposed Sipsey expansion was such an infinitesimal part of Alabama timberland, and protecting it would have insignificant negative economic impact; if even the Forest Service and the state's primary trade association for the timber industry acknowledged the need for at least some additional wilderness designations; and if local councils of government were advocating doubling or tripling the size of the Sipsey Wilderness, then why did it take a bitter campaign lasting until the late 1980s to expand the wilderness in the Bankhead National Forest?

One reason was the attitude of the Reagan administration. Reagan appointees in the Agriculture Department directed the Forest Service to recommend no RARE II "further planning" areas for immediate wilderness designation. Then, the Reagan administration formally opposed the Sipsey Wilderness expansion once it reached Congress, encouraging and justifying those in positions of power or influence who were antiwilderness.

Second, there arose local opposition in Winston County of a unique character. These people made exaggerated predictions of economic collapse, condemnation of private homes, and the like, frightening and motivating a lot of rank-and-file people employed in northwest Alabama's timber industry.

Failure during the early 1980s of Frank McAlpine's Bankhead Forest Industries and consequent closing of the mill in Grayson also reinforced local fears. All the information publicly available at the time showed conclusively that the Sipsey expansion had nothing to do with the mill's closing. Timber revenues from logging the Bankhead rose from $1.214 million in 1981 to $1.56 million in 1982 to $1.736 million in 1983, despite the proposed Sipsey expansion lands having been off limits to logging since 1977.[10] McAlpine himself blamed his financial troubles on a slump in housing construction. Moreover, the taxpayers of the state and nation, even Pres. Jimmy Carter himself, had done their share to keep the Bankhead mill in operation. In October 1980, following a personal plea by Congressman Tom Bevill to the president, the federal Small Business Administration and Farmers Home Administration approved Bankhead Forest Industries' application for a $1.5 million "lifesaving" loan, after having first rejected it as not being a "quality credit situation as required by our regulations."[11] The mill ultimately failed anyway, lending credence to doomsayers, regardless of the actual cause.

The third factor was the U.S. Forest Service's continuing attempts to retain control of the wilderness issue through their forest planning process. Back in the 1970s Congress passed the National Forest Management Act, which required the Forest Service to produce, on a regular basis and with public participation, land management plans for all the country's national forests. The early 1980s provided the first occasion for the national forests in Alabama to comply with the act. The forest supervisor's office, claiming that the draft forest plan for Alabama's national forests would be ready as early as 1982 or 1983, urged delay until the service could render a "professional" opinion.[12] As the years dragged on, this date was repeatedly pushed back, with the draft plan finally coming out in 1985. The promise of a "solution" through this plan gave opponents something to hide behind, a justification for continuing to stall congressional action.

But these were all only secondary factors. The real reason the Sipsey expansion fight dragged on year after year was the attitude of Sen. Howell T. Heflin, Democrat of Tuscumbia. As Heflin said after leaving office, "I still don't like the wilderness movement. . . . I don't want to turn a large portion of the country over to non-productive activities."[13]

One of the great lessons of the Sipsey expansion battle, and perhaps its most compelling aspect, was the way in which a person in a position of power can control the terms on which an issue is handled. By force of his office, Heflin succeeded in casting the matter as harmful to the timber industry, despite the insignificant amount of land involved. His insistence on this point angered and disappointed the conservation community and its supporters in the state media, for the facts—easily available to anyone making the effort to obtain them—were clearly to the contrary.

His legislative assistant in the early 1980s was one Charles R. "Charlie" Mitchell, a holdover from the staff of the late Sen. Jim Allen. Mitchell acknowledged to both Mike Leonard and me on separate occasions the legitimacy of the economic facts that we doggedly produced to demonstrate that the timber industry would not be hurt by the Sipsey expansion. But, Mitchell told me, "Sometimes perception is more important than the truth." Howell Heflin would have numerous opportunities over the coming years to advance his perception of the issue, and since I promised to let the senator speak for himself, the opportunity to hear him out is at hand.

CHAPTER FOURTEEN

Priceless Gift of God

EXPANDING THE SIPSEY WILDERNESS was not just an attempt to add acreage for the sake of adding acreage; it was designed to bring virtually the entire watershed of the upper West Fork Sipsey River, and the unique ecosystem it contains, under permanent statutory protection. In so doing, it would also guard the water quality and integrity of what was destined to become Alabama's only national wild and scenic river.

Imagine, if you will, the upper West Fork Sipsey's watershed as resembling a hand held upright in front of your face, fingers spread and pointing upward. The original Sipsey Wilderness of some 12,700 acres is represented by the lower half of the palm, while the arm is the river as it flows out of the wilderness, south toward Lewis Smith Reservoir. The upper half of the palm and the extended fingers represent the parts of the watershed that remained unprotected: Tedford Creek, Mattox Branch, Upper Thompson Creek, Braziel (pronounced locally as BRAZZ-el) Creek, Hagood Creek, Flannagin Creek, Upper Borden Creek, and Montgomery Creek. Expanding the wilderness to include all these "fingers"—all except Flannagin Creek, which was excluded due to large private inholdings and recent timbering activities—contained about 26,000 acres of the proposed 30,000-acre enlargement.

The remaining 4,000 acres lay to the northeast of the Sipsey Wilderness, across Alabama Highway 33, in the upper watershed of the Brushy Fork of the Sipsey River. Wild and rough as a cob, upper Brushy Fork was (and still is) a legitimate wilderness candidate in its own right. As it developed, Congressman Ronnie G. Flippo of Florence was willing to include this in his legislation as part of a single package that embraced all the RARE II lands in the Bankhead National Forest (see map p. 90). Only about 1,000 of the 30,000 total acres lay outside Flippo's Fifth Congressional District, those being along the southern boundary of the existing Sipsey Wilderness, across the

Lawrence County line in Winston County, in the District then represented by Congressman Tom Bevill of Jasper.

Flippo's receptiveness to sponsoring a Sipsey expansion began to manifest itself at about the same time that Congressman Bill Nichols was finally starting to act on the Cheaha Wilderness proposal. In May 1981 Flippo wrote, "I fully understand and appreciate the need to protect our national resources. The Bankhead Forest is a national treasure that deserves careful consideration. I would be inclined to introduce a roadless area bill at the appropriate time provided the bill would receive widespread support in the Fifth Congressional District."[1]

Prior to this, Alabama's senior senator, Howell Heflin, had written, "Let me assure you that I appreciate to the fullest the magnificent contribution which the Alabama Wilderness Coalition has made to our State . . . The establishment of the original Sipsey was indeed a giant step to preserve this priceless gift of God to all present and future Alabamians."[2]

With both Nichols and Flippo moving to produce wilderness legislation, and Heflin seemingly sympathetic, 1981 was a time of heady promise for Alabama's environmental community. Still, we had to respond to Mr. Flippo's request for a convincing expression of support within his congressional district, and here, a few key individuals stepped forward to represent northwest Alabama's interests, none more visible or colorful than Charles W. Borden.

Raised in the Bankhead Forest near Pine Torch Church and Brushy Lake, Charles and his siblings were the seventh generation of Bordens to live in the Bankhead area, their lineage proudly including Cherokee Indian blood. His family was very poor and had to subsist on welfare for a time during his childhood.

[Charles] made his own breaks in life [working] his way through college and dental school.

"I think he's a really great individual," says Moulton attorney Dave Martin. "He's a product of his environment, that's for sure. He just created his own breaks. He's done it all the hard way—honestly and unselfishly. He typifies the American dream."

To Borden, an avid hiker and hunter, the forest is as much a part of him as he is of it. When he talks about Bankhead, he sounds like an adoring parent talking about a favorite child.

"We all need areas where we can roam, where we can go over a hill to see what's on the other side, to hear the wild turkeys gobble and to hear the owls hooting," he said, " . . . a place to get away from the stresses of everyday life, to leave the rat race behind. There's good hiking and hunting in the forest, but it's also good for rejuvenating the spirit.

"People are getting more vocal," said Borden. "People in Lawrence County and Alabama are incensed about what's happening in the Bankhead. Just driving through, they see the clear-cut areas and ask, "What's happening? . . . Why can't we protect the forest?"[3]

Charles had served on the board of directors of the Alabama Conservancy in the late 1970s, playing an important role in identifying Bankhead Forest lands for the Wilderness Coalition to nominate for the Alabama RARE II inventory. Now in his early thirties, living in the Bankhead and practicing dentistry in Moulton, Borden was a tireless advocate of the wilderness expansion, haranguing virtually every Lawrence County elected official—from the district attorney to the probate judge to the superintendent of education to the tax assessor—into sending letters of support to Flippo, Heflin, and Alabama's junior U.S. senator, Republican Jeremiah Denton of Mobile.

Charles says that getting this support wasn't difficult. "Most of the public officials were hunters and fishermen and conservationists that loved the Bankhead like I did. There was a widespread sentiment to do what they could to protect the beauty of the forest."[4]

Borden's friend, young Luke Slaton, was in the process of taking over his family's publishing business in Moulton, which provided Lawrence County's only newspaper, the weekly *Moulton Advertiser*. Luke published an editorial bragging that the county "has the chance to become the doorstep to one of the largest wilderness areas in the southeast United States."[5]

In adjoining Morgan County, the *Decatur Daily* joined "with conservation leaders, state officials and concerned citizens in asking Representative Ronnie Flippo to sponsor and promote legislation to add those [30,000] acres to the wilderness program," while the *Cullman Times* said, "it sounds like a great idea that we should get cracking on. . . . [W]e also urge U.S. Rep. Ronnie Flippo to introduce appropriate legislation."[6]

Up in the Shoals Area, we found a ready ally in the mayor of Florence, J. Hollie Allen, who enlisted the support of the president of the University of North Alabama, the Alabama Mountain Lakes Tourist Association, and the local newspaper, the *Florence Times-Daily*, which called upon "our people in Washington—Flippo and Sens. Howell Heflin and Jeremiah Denton—to see that part of one [of] Alabama's most valuable resources, its untouched land, is preserved." Soon, even the very conservative *Huntsville Times* joined in editorial support, so that by late fall of 1981, Flippo felt comfortable in announcing his intention to sponsor Sipsey Wilderness expansion legislation.[7]

No one was more surprised than I when the *Birmingham News,* our nemesis in fighting the Odum Scout Trail battle, joined in editorial support for Flippo's action: "Hopefully the project will get a warm reception in Congress. It is one that could provide our state with a unique and distinctive outdoor area that many others could only envy. Flippo should be commended."[8] A *News* editorial like this begged to be investigated, so I turned to my good friend John Northrop, a former reporter for both the *News* and the *Birmingham Post-Herald.*

"Well," Northrop laughed, "a couple of things have happened since your Conservancy days. First, the *News* has a new editor. Second, my friend Ron Casey, who's been camping with me in the Sipsey several times, is on the editorial staff now. Ron wrote that editorial and he was able to get it through because he concentrated on the benefit to hunters and fishermen. Also because he published it on a Saturday when the editorial page editor has the day off!"

Ron Casey rose quickly in stature and position at the *Birmingham News,* ultimately winning a Pulitzer prize for his part in a *News* series on tax reform and becoming editorial page editor himself. Ron was the Sipsey expansion's staunchest friend in the state's daily newspaper community, relentlessly tasking Senators Heflin and Denton over the years for their attitude toward Flippo's legislation. He even apologized to me for my having to call him from time to time to ask for editorials. "I should have done that already," he'd say. Often as not, however, he'd reply, "I've already written that, it will run on Sunday." Ron Casey died of heart failure at forty-eight. He is still remembered, both in newspaper circles and out, as the model for ethical, progressive journalism in Alabama.

As for Howell Heflin, about this time he began to harden his heart toward his "priceless gift of God." Naturally, we had been sending him copies of all the letters of support and editorials that were being generated, and asked him to sponsor companion legislation in the Senate. He replied, "I would very much prefer to give [Flippo's] bill my close and careful study before considering any cosponsorship of similar legislation in the Senate, and I am sure you understand and appreciate why I feel I must resort to this course of action."[9]

Flippo's legislative assistant, Frank Toohey, felt that Heflin would come around in time and encouraged us to go after additional endorsements. One of the most important came through the state game and fish director, Charles Kelley, who had rendered such invaluable assistance in resolving the Talladega

Scenic Drive dispute. With Kelley's help, we gained the backing of the newly appointed state conservation commissioner, John M. McMillan Jr., a former state legislator. Because his background was in the timber industry, his endorsement was particularly significant: "I am in favor of rational use of our forests and recognize that the greatest percentage of our wood lands must be retained for multi-use purposes. I also recognize that the designation of the RARE II lands as additions to the Sipsey Wilderness will only represent approximately one-fifth of one percent of Alabama's forest lands; and I feel that this minimum acreage can be spared to assure that in future years we have at least part of our forests in a natural state whereby future generations can enjoy it."[10]

Figuring that there would be no harm in pressing our luck, I asked McMillan, "Commissioner, do you think that Governor James might sign a similar letter? We wouldn't need anything elaborate, just something like, 'I express my support for this proposal.'" About a week later, McMillan sent me a copy of a letter that James had sent to Howell Heflin:

Dear Senator:

It is my understanding that Congressman Ronnie Flippo plans to introduce legislation to enlarge the Sipsey Wilderness Area in the Bankhead National Forest. As you know, there is a great deal of support within the state for this proposal.

I, too, want to add my support for this effort.

Sincerely,

Fob[11]

Armed with these important expressions of support, we were able to quickly gain endorsements from Lt. Gov. George McMillan Jr., Attorney General Charles A. Graddick (whose office had challenged the relocation of the Odum Scout Trail during the Cheaha Wilderness fight), and Secretary of State (and future attorney general, lieutenant governor, and governor) Don Siegelman, while Florence mayor Hollie Allen obtained the backing of the Alabama League of Municipalities. Luke Slaton and Charles Borden got supporting resolutions from the Moulton–Lawrence County Chamber of Commerce and the Lawrence County Association of Elected Officials, as well as a letter of endorsement from the mayor of Decatur, Bill J. Dukes. More editorials and supporting statements from public figures throughout Alabama appeared on almost a daily basis.

On April 1, 1982, Ronnie Flippo introduced in the U.S. House of Representatives H.R. 6011 to enlarge the Sipsey Wilderness Area by approximately 30,000 acres.

In a prepared statement Congressman Flippo said:

I have never proposed a bill that drew more support than the Sipsey bill. . . . Practically every wildlife, conservation and environmental organization, as well as most of the newspapers and even the governor have endorsed this proposal.

There are almost 170,000 acres in the Bankhead Forest. I believe that the people of the area are entitled to a portion of that for hunting, fishing, hiking, primitive camping and other outdoor recreation activities.

This bill protects and preserves the area for our children and grandchildren.[12]

To create a bipartisan aura for the legislation, we convinced Republican U.S. representative Jack Edwards of Mobile, the hometown of Sen. Jeremiah Denton, to cosponsor the bill. In Birmingham, we turned to our area's congressman, Republican Albert Lee Smith Jr., and, largely due to the enthusiasm of one of his young staff members, outdoorsman Gilbert Johnston, Smith also agreed to be a cosponsor. Soon, Congressman Bill Nichols added his Cheaha Wilderness proposal to the legislation, making it the Alabama Wilderness Act of 1982. Hearings on the legislation before the House Subcommittee on Public Lands were to be held in May.

It seemed at the time that the Sipsey expansion was unstoppable.

CHAPTER FIFTEEN

SWUFFL

Our first indication that local opposition was organizing came in a letter to the editor in the *Moulton Advertiser:*

This is a letter concerning the Bankhead Wilderness Area. Myself and some interested persons have organized an organization to try to prevent more forest lands from being added to the present wilderness.

This not only effects [*sic*] my business but it effects your business too. Here is a conservative estimate of money lost each year, with the present amount of acres—12,726: 2,000,000 board feet × $100 equals $200,000 × 25 equals $5,000,000!! Now here is the estimate of money that will be lost if the 30,653 acres that are proposed are added: 6,000,000 board feet × $100 equals $6,000,000 × 25 equals $15,000,000. This is the total impact of forest products only! Again I say this is very conservative.

We need all the support from area citizens we can get. . . . PLEASE give us your support.

Marshell Frost, President
Society for the Wise Use of Federal Forest Lands[1]

This society, which became known by its acronym, SWUFFL, was composed primarily of loggers based in Winston County. Its president, Marshell Frost, was the owner of a small timber operation in that county, and he was motivated as much by a philosophical opposition to wilderness designations as by perceived threats to his business. The other moving force behind the group was William J. "Bill" Bustin, formerly a Bankhead National Forest district ranger, now operating a private timber consulting business in Rabbittown on the western edge of the Bankhead.

We soon learned that the mere existence of this opposing group rendered the Sipsey expansion suspect in the eyes of Alabama's delegation to the U.S.

Senate, and that the burden was not upon SWUFFL to prove the legitimacy of its claims of economic doom but upon the Wilderness Coalition to refute them.

Thus what followed was a period of three years—until it began to prove fruitless—in which the Wilderness Coalition relentlessly pursued facts to establish the lack of a threat to the local timber industry. To understand these facts is to understand why the unwillingness of Sens. Howell Heflin and Jeremiah Denton to acknowledge them and to support the Sipsey expansion produced such incredulity in Alabama's conservation community and elsewhere.

We began by emphasizing—as does this book, for it is a fact that is relevant to all of Alabama's past and future wilderness proposals—that the 30,000 acres in the Sipsey enlargement amounted to less than two-tenths of 1 percent of Alabama's 21.3 million acres of commercial forest land, and that a Sipsey Wilderness expanded to 42,700 acres, as originally proposed by Mr. Flippo, totaled only 7 percent of the national forests in Alabama.

But the crux of the opposition was that it was the local, and not necessarily the state, timber industry that would be harmed. A few inquiries on our part led us to the work of Fred Holemo and Bill McKee of the Alabama Cooperative Extension Service at Auburn University, who compiled annual reports of timber industry revenue on a county by county basis from the late 1970s through 1986. The 1981 figures, which were the most recent available at the time the first Sipsey expansion legislation was introduced, showed that the timber industry in the three Bankhead Forest counties—Lawrence, Winston, and Franklin—produced total revenue of $4.321 million.[2] Meanwhile, the U.S. Forest Service released an estimate that $196,000 of annual timber revenue would be "lost" if the additional 30,000 acres in the Sipsey achieved wilderness designation.[3] Using the Forest Services figures, then, and expressing them as a percentage of total revenue, we demonstrated that of every dollar of timber revenue produced in the three counties, the Sipsey expansion would "cost" less than a nickel, and that reducing the size of the expansion proposal would have no meaningful benefit to local industry. A 10,000-acre reduction, for example, would be "worth" less than one cent for every dollar produced in the three county area.

We also contested the accuracy of the Forest Service's estimate of $196,000 in "lost" annual timber revenue attributable to an expansion of the Sipsey. When this prediction was made public, we asked Alabama forest supervisor

Joe Brown for an explanation of how the figure had been computed. He replied that it was an average, obtained by dividing a "predicted" growth for the entire Bankhead by the total number of acres in the forest, then multiplying the result by the acreage in the Sipsey expansion. What he did not say, and what Congressman Ronnie Flippo's investigations revealed, was that the actual average timber yield per acre for the expansion lands was 40 percent lower than in the rest of the Bankhead.[4] This made sense, as the land was more rugged than in much of the forest, which is why it had retained its essential wildness. In any event, 40 percent lower meant that the loss figure should have been $117,600, not $196,000. Nevertheless, rather than get bogged down on this, we used the Forest Service figures, since even they confirmed the relative insignificance of the "cost" of expanding the Sipsey.

Acreage allocations reinforced all these conclusions. Commercial timberland in Lawrence, Winston, and Franklin Counties totaled 787,700 acres, meaning that an expanded Sipsey Wilderness of 42,700 acres was still only about 5 percent—a nickel's worth.[5]

Equally compelling were the extension service's annual timber revenue figures for public lands alone in the Bankhead Forest. In 1981 loggers produced revenue of $1.214 million from the public lands; in 1982, it rose to $1.56 million; and in 1983, to $1.736 million. These production figures were particularly significant since all 30,000 acres in the proposed Sipsey expansion had remained off limits to timbering during the entire period—in fact, since 1977, when they were "inventoried" by the RARE II process.

On May 29, 1983, Judy Johnson, special assignments editor for the *Anniston Star*, reported that the north Alabama timber industry was growing, rather than shrinking. This meant that even if in the long term a lower timber base in the national forest meant fewer jobs there, new jobs were opening up elsewhere in the immediate area. This is certainly borne out by the 2001 figures, described below.[6]

Beginning in the 1990s the job of producing the annual forestry production reports fell to the Alabama Forestry Commission, a state agency, using the same methodology employed in the 1980s by the extension service. The commission produced a report in 2001 that gives a snapshot of the most recent year in which logging was prevalent in the Bankhead; thereafter, the forest experienced a dramatic decline in commercial timber production. Challenges to management practices brought by citizen activists ultimately resulted in a significant turnover in Forest Service supervisory personnel, and reform in the agency's attitude toward logging in Alabama's small national forests. (The epilogue to this book examines these changes in detail.)

The 2001 report reflects total forestry receipts in Lawrence, Winston, and Franklin Counties of $17.131 million, up almost 300 percent from 1981. Cash receipts from logging public land alone totaled $3.167 million, more than twice the 1981 production.[7]

Another claim often put forward by Sipsey opponents was, in the words of Howell Heflin, that "there are projections that by the year 2000 there is going to be a serious shortage of lumber in America. So you've . . . got the issue of future lumber needs."[8]

This also proved to be untrue. During the fifty-year period from the early 1950s through the turn of the twenty-first century, pine plantations increased from 1 percent of total pine forest acres in the southeastern United States to 48 percent.[9] By September 2002, according to an Associated Press analysis, there was a "glut" of pine trees in the South "that will flood the market with timber over the next 15 years." Pulpwood prices were "severely depressed throughout the South, [due], in part, [to] a glut of trees from the first thinning of trees planted under" a federal program known as the Conservation Reserve Program of 1985, which "led to a profusion of pine plantations across the South."[10] Now, granted that neither Heflin nor the other opponents had any better crystal ball than anyone else, the point is that the concern about future timber needs was addressed to the point of overkill by a separate federal program specifically tailored to the situation. Defeating or limiting the enlargement of a tiny wilderness area in Alabama could never have produced such a result.

Clearly, the Sipsey Wilderness expansion has not adversely impacted the forest products industry in the least, and would not have done so even if the initial 30,000 acres had been enacted. The issue of lost timber production has always been a sham, and we said so, plainly, at every opportunity.

But these arguments were for the future; in early 1982, the Wilderness Coalition's first task was to become familiar with SWUFFL and the claims it was advancing.

Charles Borden attended the first public meeting of the group in order to get a feel for its designs, as well as to represent the Morgan County chapter of the Bowhunters of Alabama, which was supporting the wilderness expansion.

Charles was never shy about speaking his mind: "If it were left up to this society, they would clear cut it tomorrow, put it in pines and make it into a tree farm because all they're interested in is the buck. Those of you who know Bankhead, know it is being destroyed and this is the only way we can protect it."[11]

After the meeting Borden sent me copies of the materials SWUFFL handed out. They included a flyer, a four-page letter signed by Marshell Frost as president of the society, and a "fact sheet."[12]

The flyer's cover had a drawing of a triangle with a large open eye in its middle, surrounded by the words, EQUAL JUSTICE, followed by, Society for the Wise Use of Federal Forest Lands, and a contact address and telephone number. The bottom of the cover read,

WISE CONSERVATION AND USE OF PUBLIC LANDS IS A CONSTITUTIONAL RIGHT OF ALL THE PEOPLE.
WISE USE MUST BE DICTATED BY THE MAJORITY OF THOSE WHO USE THESE LANDS.
SWUFFL

The interior of the flyer contained these highlights:

Wise conservation and use of public lands is a constitutional right of all the people.

Public lands should be managed after local citizens, cities, and county governments and groups have proper notification of their ownership and the management technique and use planned. This includes proper local input into management decisions that effect [sic] taxes, property ownership income of local industry and related decisions that effect life, liberty, and the pursuit of happiness or profession guaranteed by the constitution of the United States.

Compromise is good only when it is properly applied and granted by the citizens or their representatives.

We believe the National Forests should be managed for more than one use . . . each being limited to the economic need of local citizens first.

NO FURTHER RESERVATIONS SHOULD BE MADE IN THE WILLIAM B. BANKHEAD NATIONAL FOREST FOR WILDERNESS AND THE PRESENT RARE II STUDY AREAS SHOULD BE RELEASED FOR GENERAL FOREST MANAGEMENT AND USE.

108   *Sipsey Wilderness Expansion*

Disadvantages of Wilderness

Less than 1% of the citizens will use the areas.

Citizens access to wilderness will be allowed only by foot. No horses will be allowed or motorized vehicles.

Private landowners within the wilderness may have their land CONDEMNED, PURCHASED, EXCHANGED or otherwise disposed of if they do not agree to comply with Wilderness laws and regulations.

Marshell Frost's accompanying letter contains many assertions, among them:

We are opposing [Flippo's] bill because the economic IMPACT will be a DISASTER TO THE ARES [sic] IN AND AROUND THE FOREST.

The Forest Service should professionally recommend any addition to the present Wilderness system. . . . Arbitrary selection of 30,653 acres is not *in the best interest of the majority of the people or professionally or economically wise.*

Research shows wilderness visitors to be:
   1. Highly educated
   2. Above average income
   3. White collar workers

[Private] land can be taken and condemned by the Federal Government.

Can the Hikers Club, Sierra Club, Audubon Society, and Alabama Conservancy find jobs for the unemployed people that will be left without a job if this proposal is passed? Can these various clubs that are supporting the wilderness find homes for the landowners in the Forest? . . . REMEMBER IF THIS BILL IS PASSED DO NOT BE PROTESTING WHEN ANY PAPER PRODUCTS PRICES SKY ROCKETS!!!!! IF YOU DO NOT EXPRESS YOUR VIEWS AND CONTACT YOUR LOCAL CONGRESSMAN WILL YOU BE RESPONSIBLE FOR PUTTING A CITIZEN THAT HAS EARNED HIS LIVING FROM THE FOREST OUT OF A JOB??????????????

Finally, the fact sheet repeated several claims in the other handouts, and added a few more:

The total annual economic impact because of the present 12,726 acre wilderness is $5,000,000 and [an]other $10,000,000 would be lost if another 30,653 acres is added.

The proposed selection process is not contiguous; it leaves gaps in the selected area and encompasses many acres of private lands including lands where people are now living.

If the proposed wilderness area is passed, private landowners would have one of four choices:
(1) Sell land to Forest Service.
(2) Exchange the land for acreage outside the wilderness area.
(3) The land can be used in the same manner that it was used before the wilderness was proposed.
(4) The land can be condemned and taken by the Federal Government.

Roads that have been used by forest visitors for years may be closed.

Pine Beetle infestation cannot be controlled and damaged timber cannot be harvested by Forest Service inside wilderness areas.

The source of the "facts" contained in these documents, which so agitated Marshell Frost and the loggers who populated SWUFFL, was former Bankhead ranger Bill Bustin.

Charles Borden says, "I resented Bill Bustin the most. I had always had pretty great respect for the Forest Service. I thought that they were the stewards of our National Forests, supposedly protecting this great public resource for us. But here was Bustin, a former district ranger, organizing opposition to protection of the forest, and using a lot of misinformation to do it."

However, Borden describes the rank-and-file SWUFFLites as "sincere people. Most of them came from Winston County, which is very poor and the people there have almost nothing but logging as a way to make a living. They saw this wilderness proposal as a matter of survival. It was easy for Bill Bustin to upset them. I knew that they were being misled so I never had any personal animosity towards them, even though some of them threatened me and my family."

For the Wilderness Coalition, the challenge was to refute SWUFFL in a simple and clear manner so that it could be understood by the press and—we hoped—by Senators Heflin and Denton and their respective staffs.

Putting aside SWUFFL's manufacture of new constitutional rights of "wise use of public lands" and "pursuit of profession," the most troubling of its contentions was that expanding the Sipsey Wilderness would produce an economic "loss" of $10 million dollars.

SWUFFL's calculations began with the premise that the 30,000 acres contained an annual yield of 4 million board feet of timber, that is, that the timber industry could count on 4 million board feet being available for harvest year after year after year, and, of course, that there would be demand for that level of production during all that time. To that figure, they assumed a uniform annual price of $100 per thousand board feet, producing a direct "loss" of $400,000. Then, they claimed that each of those dollars produced other economic benefit ("Labor, gas, oil, Social Security taxes, workmen's compensation, S.U.I., Support organizations, and people to get the product to its final use") at a "conversion factor" of 25 to 1. Thus, $400,000 × 25 was $10,000,000.

The simplest retort to this came from the fact sheet that the Forest Service's Montgomery office produced with respect to Flippo's legislation, which stated that annual timber revenue attributable to the Sipsey Expansion lands was $196,000, not the $400,000 claimed by SWUFFL.[13]

According to the Alabama Cooperative Extension Service at Auburn University, loggers produced gross revenue of $1.214 million from the public lands in the Bankhead Forest in 1981. Total Bankhead acreage was, in round figures, 180,000, of which the 30,000 acres in Flippo's bill constituted 17 percent. The Forest Service's estimate of $196,000 was a little more than 16 percent of the $1.214 million in gross timber revenues, making it at least bear some relationship to the amount of acreage involved. SWUFFL's $400,000, however, was nearly 33 percent of 1981's gross revenues, meaning that SWUFFL was claiming that 17 percent of the land in the Bankhead—despite its ruggedness and relative inaccessibility—would produce almost 33 percent of gross revenues on an annual basis, an assertion that was absurd on its face.[14]

As for the 25 to 1 "conversion factor" claimed by SWUFFL, this was basically irrelevant. If one is to apply the conversion factor to the supposed loss, then it must also be applied to the total production figure to put it in perspective. In other words, if the Forest Service's $196,000 is multiplied by 25 ($4,900,000), and the result divided by 1981's gross timber revenue of $1.214 million times 25 ($30,350,000), the loss is still only about 16 percent of total revenue.

However, the whole idea that the local timber industry would inevitably suffer any loss at all from an expansion of the Sipsey defied common sense, as it was based on the assumption that the Bankhead National Forest was the only source of commercial timber in the area. In fact, the three Bankhead Counties—Lawrence, Winston, and Franklin—contained 787,700 acres of commercial timberland, meaning that expanding the Sipsey to 42,700 acres would still leave 745,000 acres, or about 95 percent, of the area's forests available for development. In that three-county region, loggers had produced gross revenue of $4.321 million in 1981. The Forest Service's $196,000 was only 4.5 percent of those actual revenues, not a nickel's worth out of every dollar.

As for the Bankhead National Forest itself, gross timber revenue from logging went from $1.214 million in 1981 to $1.56 million in 1982 to $1.736 million in 1983, even though all 30,000 acres in the Sipsey Expansion bill had been out of production since 1977.

In short, "harm" to the timber industry was a matter of perception, not of fact.

The other big issue raised by SWUFFL was, of course, that of condemnation. As with the Cheaha Wilderness, this problem was solved at the outset, first, by gerrymandering out all the private lands on the periphery of the RARE II areas and, second, by having the legislation designate the new areas as additions under the original 1964 Wilderness Act, which extends no right of condemnation to the government whatsoever. These facts did not prevent SWUFFL from using the issue to stir up local hysteria, however.

As for SWUFFL's other claims, we anticipated several (although we chose to let some speak for themselves, as I do now). On the issue of visitation, for example, we had already provided both senators copies of Forest Service estimates that the use of the existing 12,700 acre Sipsey Wilderness Area rose from 8,500 visitor days in FY 1977 to 11,600 visitor days in FY 1981, a 36.5 percent increase in just four years.[15]

The issue of road closures gave us the opportunity to be creative. Two east-west Forest Service dirt roads and one north-south road ran through the middle of the proposed expansion. One of these, Forest Road 208, known as the Northwest Road, ran the entire width of the area, east to west. The acreage in the original Sipsey Wilderness together with the 7,000 acres the Forest Service had recommended for wilderness designation in RARE II lay to the south of the Northwest Road, while the remainder lay to the north (see map p. 90). Flippo's proposal, ultimately adopted by the House of Representatives, was to leave the Northwest Road open, "to enhance public use of the wilderness.... In particular the elderly, handicapped, and the young would benefit from the continued use of this road," said the congressman. "It would provide these individuals with easy access to this rugged area and to the very popular Bee Branch section of the existing Sipsey Wilderness."[16]

Flippo also proposed, and the House agreed, that the expansion be open for horseback riding, as, indeed, most National Forest Wilderness Areas are. The original Sipsey Wilderness, due to its small size, and because it contained the deepest cuts of the watershed's canyons, sheltering much of its unique flora, had been considered by the Forest Service as inappropriate for the heavy impact of horses and had been closed to riding. The expansion area would be open to fill that need.

We felt that we, and Congressman Flippo, had been innovative in crafting a wilderness proposal that respected the needs of many different types of forest users, while still preserving the West Fork Sipsey watershed for its wilderness values. And the proposal gained wide acceptance, except in Alabama's U.S. Senate delegation, where the only thing that seemed to matter, regardless of party affiliation, was the perceived impact upon the timber industry.

Blanche Dean (*standing, left*) and Mary Burks (*seated, left*), on a field trip with friends, April 1968. Photo by Harriett H. Wright.

Sen. James B. Allen Jr., circa 1975. Courtesy *Birmingham News*.

Sen. John Sparkman, circa 1979. Courtesy *Birmingham News*.

Rep. Bill Nichols (*left center*) and environmental activist Mike Leonard, with color guard of Boy and Girl Scouts, unveil Cheaha "Wildernes" marker at the Cheaha Wilderness dedication ceremony, August 12, 1983. The "Wildernes" misspelling was later corrected. Photo by author.

Rep. Bill Nichols (*left center*) and Mike Leonard, Scoutmaster Moran Coburn, and color guard of Scouts from Troop 124, Talladega School for the Deaf, at the Cheaha Wilderness dedication ceremony, August 12, 1983. Photo by author.

Sen. Jeremiah Denton, circa 1986. Courtesy *Birmingham News*.

Undated photograph of Sen. Howell T. Heflin (*right*) and constituent. Courtesy *Birmingham News*.

Sipsey Wilderness expansion dedication ceremony, April 14, 1989. *Left to right:* Rep. Ronnie G. Flippo, Mary Burks, the author, Charles Borden. Photo by Virginia Randolph.

The author (*right*) presents a plaque of appreciation to Rep. Ronnie G. Flippo at the Sipsey Wilderness expansion dedication ceremony, April 14, 1989. Photo by Virginia Randolph.

Pete Conroy, ca. 2004. Photo by Stephen Gross.

Francine and Bruce Hutchinson, ca. 2004. Photo by David W. Hutchinson.

Rep. Bob Riley speaking at the Dugger Mountain Wilderness dedication ceremony, April 1, 2000. Photo by Carol Moore, Jacksonville State University.

Sen. Jeff Sessions speaking at the Dugger Mountain Wilderness dedication ceremony, April 1, 2000. Photo by Carol Moore, Jacksonville State University.

CHAPTER SIXTEEN

Gutless Sellout to the Timber Industry

SWUFFL GENERATED A LOT OF HEAT, and it was interesting to witness how various politicians handled it. Some of them got so many letters from Winston County loggers that they developed a form letter in response—and sent it to anyone who wrote, pro or con.

Governor James's approach was to pass the buck: "I have received your letter opposing any addition to the Sipsey River Wilderness. It is my understanding that the establishment of a wilderness area or the enlargement of a wilderness area on U.S. Forest Service lands will require favorable action of Congress. It is also my understanding that the proposal for approximately 12,000 acres [sic] additional to the Sipsey area has been presented for consideration and I do not know of any firm action having been taken as of this date. Perhaps you desire to contact your Congressman relating to this matter."[1]

Birmingham congressman Albert Lee Smith, a cosponsor of the Sipsey bill, chose to obfuscate: "I am very concerned that our approach to environmental protection in Congress must be balanced with economic protection. We must balance wilderness preservation with multiple use forests and industrial development. We must make these decisions with a view to our state as a whole and our nation as a whole. . . . I hope you will continue to express your views to me."[2]

Montgomery's pistol-packing Republican mayor Emory Folmar was not intimidated. When we first approached him for an endorsement of the Sipsey expansion, he greeted us with an enthusiastic, "Those loggers in Winston County are just damned greedy!" Now, he was preparing to run for governor and had become a SWUFFL target. His letter read, "Thank you for taking the time to write me regarding my support of the Sipsey Wilderness Bill in the Bankhead National Forest. . . . My support comes from the fact that I believe this to be best for all of Alabama and for all Alabamians. There are many in opposition to this and they are quite vocal about the fact that they are going to be against me because of this. So be it. I think Alabamians should protect this part of our heritage. I also happen to think that it will be a tourist attraction for years and years to come, and I will work towards that end."[3]

But most instructive was the way Congressman Tom Bevill handled the uproar. "Special Interest Group Forces Reduction of Sipsey Acreage," read the *Florence Times-Daily* headline. "Mr. Bevill asked that the land in his district be taken out, and we are doing it," said a Flippo spokesman. Bevill's spokesman said, "The people in Winston County are opposed to this bill and those are the ones he represents."[4]

Actually, we were relieved. While the result was about a 1,000-acre reduction in Flippo's proposal, only a small corner adjacent to the Sipsey's southeastern border in Winston County was of particular concern to us, and if losing this was all that was needed to keep Bevill neutral, then we, and Ronnie Flippo, were more than willing to make the adjustment. Tom Bevill was a senior and powerful member of Congress. He could have made a great deal of trouble for Flippo and his legislation, had he chosen to do so. Instead, he accepted this token acreage reduction and let Flippo proceed otherwise unimpeded, a gesture of collegiality that we hoped might be an example for our Senate delegation.

Howell Heflin had other thoughts. The Alabama Forestry Association, the state's powerful timber industry trade association, had just passed a resolution opposing Flippo's bill.[5] Shortly before House hearings on the legislation, Heflin wrote to me,

> I may say that I have . . . received quite a bit of correspondence, both pro and con relative to H.R. 6011. I am sure that the hearings on the bill . . . should help to define many of the issues in dispute and, hopefully, resolve these issues and other questions which may arise regarding the bill.
>
> I am sure you recognize and agree that my stance with respect to H.R. 6011 is somewhat different than that of Congressman Flippo, inasmuch as my constituency is state-wide. I, therefore, feel that in fairness to all concerned, I should not make any decision at this time regarding the introduction of companion legislation in the Senate until after the bill has been publicly aired through the committee process in the House of Representatives and I have had an opportunity to thoroughly study the testimony given both for and against the legislation. . . . I just feel that in view of the intense interest which has been evidenced to me on both sides of the legislation, I should, in

the interest of fair representation wait for House action on the bill. I am sure you understand and appreciate my position in the matter.[6]

Jeremiah Denton, on the other hand, chose the tactic that he would continue to rely upon for years—wait for the Alabama forest plan, let the Forest Service decide:

> [RARE II] gave the U.S. Forest Service until 1985 to complete the study and make recommendations for the final designation of the areas that were to be given "further planning." . . . [T]hose areas must be managed as wilderness areas until the final designation is approved by Congress.
>
> The Forest Service is currently evaluating the "further planning" areas in Alabama, and it is scheduled to complete the draft forest plan and publish it for public comment before the end of this year [1982]. I believe that it would be in the best interests of Alabamians to have the benefit of the Forest Service professional evaluation before deciding on specific legislative initiatives.[7]

The Reagan administration gave Denton cover by opposing Flippo's bill when it came before the House Subcommittee on Public Lands and National Parks on May 24, 1982:

> The Reagan Administration yesterday objected to a bill that would expand the Sipsey Wilderness in Lawrence County. . . .
>
> R. Max Peterson, chief of the U.S. Forest Service, told the House subcommittee on public lands and national parks that Congress should not adopt the addition to the existing 12,000 acre Sipsey Wilderness proposed by Rep. Ronnie Flippo, D-Florence. Peterson said "further study" is needed to determine if the value of the timber on the land outweighs its value as a wilderness area.
>
> The Forest Service chief said Alabama's national forest plan is scheduled for completion late next year [1983]. In the interim, he said, the areas named in Flippo's bill will be managed to preserve their wilderness potential.[8]

I had little patience with the notion of turning the fate of the wilderness over to the Forest Service, as my response indicates: "In the final analysis, the question before the land management plan would be whether to leave either 99.9 percent or 99.7 percent of our state's forests available for commercial development. Why are the people of Alabama, through their elected representatives, not competent to resolve that question? Why is this Congress not competent to provide the answer?"⁹

Apropos of the Reagan administration's position was the *Birmingham Post-Herald*'s decision to run, in the same issue as its report on the subcommittee hearings, an editorial cartoon depicting Interior Secretary James Watt standing on a stump, ax in hand, with an accompanying poem:

Trees
by James Watt

I think that I shall never see
An environmentalist lovely as a tree
A tree under which we find
Land which shortly will be mined
A tree which does no one no good
Unless it's harvested for wood
A tree that will by summer be
A redwood deck for you and me
A tree that some young eco-freak
For his children wants to keep
Policy is made by fools like me
Not even God can save this tree.¹⁰

Despite the opposition, the House subcommittee hearings were successful. We felt we had a good delegation to support Congressman Flippo. In addition to Mike Leonard and me were Charles Borden, Doug Phillips of the University of Alabama, Hollie Allen (one-time mayor of Florence), and Luke Slaton of Moulton, whose testimony emphasized his roles as a businessman, past president of the local chamber of commerce, and editor of the *Moulton Advertiser*:

> The people of the business community welcome the addition to the wilderness area and have signified so by passing a Chamber resolution to that effect.
> As editor of the county's only newspaper, I'm in a good position to gauge the feelings of the people of the county. We believe in taking editorial stands on issues

that concern our county, and we have heartily endorsed the wilderness addition. The response from our readers has been favorable.

That's not to say that the majority of Lawrence County's citizens are avid outdoorsmen who will swarm into the wilderness area. But the people of the county are proud people. They tend to favor things that make Lawrence County unique; that give us something special. The larger wilderness area will give Lawrence County a unique attraction.[11]

Speaking in opposition, besides the chief of the U.S. Forest Service, were Bill Bustin of SWUFFL, and a new face, Leo Yambrek of Killen, who intended to run for Congress as a Republican against Ronnie Flippo later that year. Yambrek said, "Can the people of north Alabama afford to have this much timberland permanently removed from productive use for the benefit of a few special interest groups?"[12]

Being in Washington gave us the opportunity to meet with the two senators, except that Denton would have none of us. "You're not alone," one of Heflin's staff people told us. "Denton refuses to see just about everyone."

Heflin was not unfriendly, but neither was he encouraging. In fact, he wanted to change the subject. He jumped on us about the Alabama Conservancy's years of opposition to the Tennessee-Tombigbee Waterway, which had long been a pet pork-barrel project of successive members of the Alabama delegation, including Sen. Jim Allen, Congressman Tom Bevill, and now, Heflin, who had just succeeded in getting final funding to complete the big ditch. Heflin also wanted us to give him credit for legislation he sponsored requiring double liners at the hazardous waste landfill in Emelle. As for the Sipsey expansion, he said, "I've just been too busy to focus on it. Once it passes the House and comes over here, I'll take a look at it."

Following the acreage reductions requested by Tom Bevill and the Forest Service's latest computations of the acreage involved, the Sipsey expansion was now approximately 28,500 acres. It, and Congressman Bill Nichols's Cheaha Wilderness proposal, were combined as H.R. 6011, the Alabama Wilderness Act of 1982, which was reported out to the House by the full Committee on Interior and Insular Affairs in July 1982.

Shortly before that, SWUFFL managed to attract the attention of two north Alabama television stations, including one in Birmingham, by claiming that farmers whose private land adjoined the proposed wilderness in upper Borden Creek would be under threat of condemnation if the expansion bill passed. The Birmingham station sent out its news chopper to record the farmers' plight. When I learned of this, I called the news director of the Birmingham station and explained to him that the private lands were on the periphery, that the wilderness would contain only publicly owned land, and that H.R. 6011 had no right of condemnation in it. "You mean we've been had," he replied. Of course, he let me go on the air to rebut SWUFFL's charges, but the damage had been done.

We complained bitterly to both senators' staffs, and for good reason. Bill Bustin and an attorney retained by SWUFFL had attended the House subcommittee hearings and had ample opportunity to see the maps on file with the subcommittee. Congressman Flippo gave a detailed explanation of this issue as a part of his testimony. The SWUFFL representatives knew—or should have known, if they had been paying attention—that the farmers' lands were not threatened.

Did our complaints do any good? SWUFFL never raised the issue of condemnation again—publicly, at least.

On August 4, 1982, the Alabama Wilderness Act overwhelmingly passed the U.S. House of Representatives—but only after a roll call vote was forced by the Reagan administration:

Rep. Ronnie Flippo, D-Florence, and Rep. Bill Nichols, D-Sylacauga—sponsors of the bill—learned only minutes before the 349–59 vote was taken that the opposition was coming directly from the White House.

An administration policy statement was quietly circulated among Republicans last Friday asking them to vote down the measure because part of the planned expansion to the Sipsey Wilderness includes valuable timber land.

"I was a little surprised at the opposition by the administration coming so late," Flippo said. . . . "The arguments raised by the special interests over this bill do not hold water. . . . There is no shortage of timber in Alabama. Two-thirds of the state [is] covered with timber, and this land represents less than two-tenths of one percent of the commercial forest land in the state."[13]

House passage was an important milestone, but time was running out on the legislation, as 1982 was an election year and Congress planned to adjourn on October 8 so that its members could concentrate on getting reelected. Especially critical was the fact that 1982 was the end of the 97th Congress, meaning that unless there was a lame duck session following the elections, a refusal by Alabama's senators to quickly move the bill would kill it, forcing Flippo to start over again—assuming he was reelected—when the 98th Congress convened in 1983.

Heflin wrote me again to say that he had been "too busy" to make up his mind about the Sipsey expansion, but he did promise he would try to get a "speedy" hearing before the appropriate committee. But, he said, "I feel I must . . . point out that it is the Republicans and the Republican Senators who are Committee Chairmen who control the hearing agendas of the Committees of the Senate. The Democrats have no authority whatsoever in the scheduling of hearings on the legislation."[14]

Shortly, the Republican in our Senate delegation made his wishes known. In September, Jeremiah Denton announced that he was asking the Senate leadership to separate the Sipsey from the Wilderness bill and allow only the Cheaha to go forward. The move made both Flippo and Bill Nichols mad.

"We were not contacted," said Nichols's press secretary. "[H]e took this action without even the courtesy of a phone call to me," complained Flippo. "The senator's action will probably kill the bill at this late date, with only one month remaining in Congress. If this is not his intent, his action may kill the bill whether it is his intent or not. . . .

[D]elegation sources called [Denton's] move political sniping. They believe Denton is trying to kill the bill in an effort to build support for Flippo's Republican opponent, Leopold Yambrek, who has campaigned against the Sipsey expansion.

Denton also is believed to be following the lead of President Reagan, who orchestrated last-minute House opposition to the bill by circulating an administrative policy statement among Republicans.

"He's also being a chicken," said one Alabama House aide. "He doesn't want to tee off the timber people."[15]

SWUFFL was ecstatic. Ruth Ann Howell, the wife of a timber company executive in Haleyville, told the press that she "recently had a conversation with Denton concerning the bill. 'I thanked him for what he did,' she said. 'We humbly request and petition Sen. Howell Heflin and Sen. Jeremiah Denton to kill the bill.' Added Marshell Frost, 'We don't need no more wilderness, we've got 12,900 acres.'"[16]

To celebrate, SWUFFL organized a couple of media events, rolling about twenty-five logging trucks into Florence with a man dressed as a pine beetle and placards reading "Take wilderness and shove it—we need jobs."[17] Ruth Ann Howell led a group of women carrying wooden snakes with bundles of pine seedlings to deliver to Flippo's district office so that "he can plant his own forest."[18] One event was also a campaign rally for Leo Yambrek, now officially the Republican nominee for Flippo's House seat. Yambrek, who was on record as believing it unconstitutional for the federal government to own land, said that the loggers were "the true environmentalists . . . out there protecting our forests. We don't need the Sipsey Wilderness."[19]

Now, Denton began catching heat from wilderness supporters, particularly newspaper editorial boards, so that when it became clear that Congress would, in fact, reconvene after the November elections in lame duck session, Denton arranged for H.R. 6011 to be heard before a subcommittee of the Senate Energy and Natural Resources Committee. Said Denton:

> Some individuals, who seek political advantage from the situation, have accused me of trying to "kill the Sipsey proposal." Nothing could be farther from the truth.
> Although the Cheaha portion is non-controversial, legitimate questions have been raised about the impact of the proposed Sipsey Wilderness expansion. I do not believe the Cheaha measure should be held up until the questions about the Sipsey proposal have been answered.
> This bill is a matter of great interest to the people of the areas around the Bankhead National Forest . . . and to all Alabamians. I will keep an open mind on the Sipsey portion until the Forest Service has completed its study [the Alabama National Forest Management Plan] and the Energy and Natural Resources Committee has held hearings.[20]

We were unable to identify any "individuals who seek political advantage from the situation" other than Leo Yambrek, and on that score, Flippo was reelected with 82 percent of the vote in his district, and 95 percent of the vote in Lawrence County, where the Sipsey expansion lands were located. "Our two senators are not responding to the mandate of the people of Alabama," complained the *Florence Times-Daily*.[21]

Through all this, Charles Borden had been fuming, and now he decided to make public his "outrage against the gutless sell-out by Denton and Heflin to the timber industry." In letters to various newspapers around the state, he charged that

[n]either Denton nor Heflin had the backbone to stand up to the timber lobby. The green they are seeing is not that of our forest but that of money from their high-powered buddies in the timber industry.

If Heflin and Denton do not see fit to redeem themselves . . . then I would suggest that they had better start looking for a job with the timber industry. The people of Alabama will see that Heflin and Denton join the list of those unemployed.[22]

Obviously, this was not well received in Washington, and it marked the beginning of the end of Charles's effectiveness in lobbying for the Sipsey expansion; but Borden being Borden, he would do as he pleased, and later, he would please to do even more.

I was in no position to criticize, as I shared Borden's anger over the senators' intransigence and was guilty of saying to the press such things as "those who would have us believe that the Sipsey expansion will harm the state's timber industry are just not facing the truth," and "the issue of lost jobs is a sham."[23] This did not sit well in Washington, either, but it did resonate elsewhere, as we had the facts to back it up.

The hearings that Senator Denton arranged before the Public Lands and Reserved Water Subcommittee of the Senate Energy Committee were held on November 30, 1982, and amounted to little more than public posturing, with SWUFFLites making such charges as "these radical environmentalists [want] to use Bankhead National Forest as a playground for their whims and misconceptions while there are those people without jobs,"[24] and "only about 1 percent of the people would use a wilderness area . . . and the other 99 percent would suffer."[25]

Ronnie Flippo was invited to "defend" his legislation, but declined. "He's taking the position that he's passed a bill in the House," said Pete Kelley, Flippo's press aide. "His position is abundantly clear and now the Senate will have to take a position."[26]

We radicals brought another good delegation to testify in favor of the enlargement. In addition to most of us who had testified before the House, notably Luke Slaton on behalf of the Moulton–Lawrence County Chamber of Commerce, we had Congressman Albert Lee Smith Jr. of Birmingham; Huntsville native Jim Price, now southeast regional representative of the Sierra Club; state representative Roger Dutton of the Seventh Legislative District (Lawrence and Morgan Counties); and Bob Barker of the national Bass Anglers Sportsman Society (BASS). Nothing any of us had to say, however, carried any weight in the face of opposition from the U.S. Forest Service and Sen. Jeremiah Denton.

Lamar Beasley, deputy chief of the U.S. Forest Service, and Senator Denton both repeated their insistence on waiting for the Alabama National Forest Management Plan, still claiming that it would be completed in 1983. "I am not opposed to the Sipsey Wilderness nor to expanding it in some reasonable way," said Denton. "I believe that we must have further study before proceeding."[27]

Senator Heflin, however, was the one who defined the terms upon which the expansion fight would proceed in the coming years, and in doing so, coined its most memorable phrase. Admitting that "[f]rankly I do not know how many jobs could either be lost or created as a result of the Sipsey expansion," he said that there is a "great deal of controversy" about how much land should be added to the existing wilderness. "Heflin said that he supports the idea of wilderness areas, which he called 'gymnasiums of nature' [but that] the Senate must pay close attention to the economic impact of the Sipsey expansion." He expressed his belief that the Sipsey should contain "enough wilderness to adequately preserve a comprehensive showplace and gymnasium, but no more."[28]

With that, the Sipsey Wilderness expansion died in the 97th Congress.

CHAPTER SEVENTEEN

Elitist Birdwatchers and Backpackers

THE DEATH OF THE SIPSEY legislation was a bitter disappointment little relieved by the successful establishment of the Cheaha Wilderness, even for those of us who had fought the Odum Scout Trail battle for so long. In an editorial published at the conclusion of the 97th Congress, the *Birmingham Post-Herald*, after thanking Bill Nichols and Jeremiah Denton for their Cheaha work, characterized the result as "less than half a loaf":

[W]e would be remiss in not regretting the actions of Sen. Denton, and to a lesser degree, those of Sen. Howell Heflin that, at the very least postponed expansion of the existing Sipsey Wilderness . . . [which] will have to start through the legislative mill again when the Congress convenes [in 1983]. . . .

Unfortunately, the Sipsey proposal . . . has a few opponents, primarily logging interests. So when the [House-passed] legislation reached the Senate, Denton successfully separated the [Cheaha portion] . . . arguing that the U.S. Forest Service needed time to complete a study of the Sipsey area. Heflin also expressed reluctance to support the Sipsey expansion, indicating that he wanted to reduce the area involved as a compromise.

Considering the record of the Forest Service in regard to wilderness designations in Alabama, we have little faith in their land management plan, which won't be ready until late next year [1983]. The service opposed the original Sipsey Wilderness Area and for many years was opposed to the Cheaha designation. This despite the clear need for protection of the areas and the support the designations had from Alabamians across the political spectrum.

As for the loggers' arguments about the economic losses from expanding the Sipsey Wilderness, they are overblown to put it mildly. The expanded Sipsey Wilderness would represent two-tenths of one percent of all commercial timber land in Alabama. . . .

But even a small, short-term economic loss would be acceptable for preservation of this area. Once it is destroyed for immediate gains, it is gone.

The Sipsey expansion should be a first order of business for Alabama's congressional delegation when the new Congress convenes.

The Cheaha Wilderness, welcome as it is, represents less than half a loaf.[1]

Ronnie Flippo promised to reintroduce his bill, but he and his legislative assistant, Frank Toohey, urged us to see if there was some room for compromise that might attract Heflin's support. As loathe as we were to give an inch, they had a point.

It is not unusual—in fact it is probably the norm—for there to be differences of opinion between members of the two houses of Congress. The legislative process works when those members get their separate proposals enacted by their respective houses, and then try to resolve the differences, in the case of major legislation, through a formal House-Senate conference committee, or in the case of "local" legislation like the Sipsey expansion, through direct negotiations between the two sponsors. The key is to get the opposing member to take a stand.

Frank Toohey would tell me countless times over the next six years, "John, you've simply got to find something that Heflin will support, something that he'll sponsor and get the Senate to pass, and then let Mr. Flippo do his best for you."

The trouble was, Heflin seemed to have little interest in the legislation, and for now was content to lay the blame for its failure on the Republicans. As he wrote in letters to public officials who had contacted him in support of the Sipsey expansion: "When this [1982] legislation reached the Senate, however, the Republican-controlled Senate Committee on Energy and Natural Resources refused to approve the Sipsey proposal due largely to objections lodged by the U.S. Forest Service on behalf of the Reagan Administration."[2] And, "As you no doubt recall, the House of Representatives passed the Sipsey expansion proposal last year, but the Republican-dominated Senate Energy and Natural Resources Committee did not act favorably on the measure prior to the adjournment of the 97th Congress in December, 1982."[3]

Our best hope of getting any concession at all out of the timber industry lay not with SWUFFL but with the Alabama Forestry Association (AFA), which had proposed back in 1978 that we join in a committee to make recommendations on RARE II to Alabama's congressional delegation. Unfortunately, nothing ever came of the idea, as it was overruled by the AFA's full board of directors; then, in 1982, the AFA Board adopted a resolution opposing Flippo's legislation.

Nevertheless, we still maintained occasional contact with one of the individuals who had first approached us, William F. Chestnutt of Union Camp Corporation in Wetumpka. I asked Chestnutt if he would test the waters for compromise within the ranks of the timber industry, and in January 1983 he gave me his frank assessment, which I passed on to Congressman Flippo.

Chestnutt obtained Alabama national forest supervisor Joe Brown's tacit acknowledgement that the upcoming forest plan would not recommend a Sipsey enlargement beyond the 7,000 acres originally proposed by the agency at the conclusion of RARE II. Under its current leadership, Chestnutt said, the AFA was unwilling to do anything other than what the Forest Service would recommend. And Bill Bustin told him that SWUFFL was unwilling to consider any compromise.

None of this was particularly surprising. Why would the timber industry compromise as long as Jeremiah Denton was following the Forest Service's lead and Howell Heflin was doing nothing at all? However, a recent development did offer some promise. George Wallace had been elected to succeed Fob James as governor, and James's conservation commissioner, John McMillan, who had publicly endorsed Flippo's 1982 legislation, was going to work for the AFA. In the near future, McMillan was due to assume the association's chief of staff post, replacing one J. Hilton Watson, an ardent wilderness opponent. If Flippo would stick to his guns, in time perhaps John McMillan might influence a change in the AFA's position.

Armed with endorsements from the new state administration—Gov. George Wallace himself and his new conservation commissioner, John W. "Toppy" Hodnett—Ronnie Flippo reintroduced his Sipsey expansion legislation in the U.S. House of Representatives on April 12, 1983. Cosponsors of H.R. 2477, the Sipsey Wilderness Additions Act of 1983, were Congressmen Ben Erdreich, D-Birmingham (who had defeated Republican Albert Lee Smith in his reelection bid); Bill Nichols, D-Sylacauga; Richard Shelby, D-Tuscaloosa; Bill Dickinson, R-Montgomery; and Jack Edwards, R-Mobile—in short, the entire Alabama House delegation except Tom Bevill. As a result of updated acreage computations by the Forest Service, the proposed expansion totaled 27,865 acres.

Hearings before the House Subcommittee on Public Lands and National Parks were held on April 25, 1983, and were routinely concluded. Ronnie Flippo appeared, as did I, along with Ed McMahon, an Alabama native living in Washington, who gave a supporting statement on behalf of the Alabama Conservancy.

The only opposition was from the U.S. Forest Service. Forest Service chief Max Peterson again appeared, again advocated waiting for the Alabama forest plan. While admitting that the area had "high wilderness values," he claimed that it also had "high other values including timber . . . and we would like to have the information that would come from the forest planning process before action is taken."[4] I was growing weary of this and was still a little angry over last year's result, as my response indicates: "Those who claim that we cannot expand the Sipsey Wilderness until we have the permission of the U.S. Forest Service are saying in effect that Alabamians are incompetent to decide the fate of their own land. . . . The timber issue raised in opposition to the Sipsey enlargement is just as big a sham this year as it was in 1982."[5]

That Chief Peterson seemed to be recommending against wilderness designation even for the 7,000 acres the Forest Service had proposed for wilderness in RARE II caught the attention of the subcommittee chairman, Congressman John Seiberling of Ohio. Peterson replied that there was a 165-acre private inholding that had been on the northern periphery of the original Sipsey Wilderness, but was now surrounded by the 7,000 acres. The Forest Service was negotiating a swap for land outside the wilderness. "[W]e are not changing our recommendation [that it] should be wilderness at all," said Peterson. "We are just suggesting that there might be some merit to waiting until we resolve this [exchange]."[6]

Seiberling: So we are going to hold this whole thing up over 165 acres? . . . I don't think that a 165-acre inholding should hold up what is otherwise a meritorious proposal.

Mr. Flippo: Mr. Chairman, the company that owns that particular piece of land . . . does not deem the timber to be of marketable value at this time and they have already made arrangements to swap that land. . . . [I]t has been our intent not to take anyone's land who didn't want their land included in the provision. We didn't include eminent domain in the act. . . .

Mr. Seiberling: Under existing law [the owners] would have a right of access even if we put it in wilderness . . . [but the Forest Service is] concerned that until they worked [the exchange] out they would like the maximum flexibility and that is one reason they don't want to put the area in wilderness.

Mr. Flippo: Mr. Chairman, I am under the impression that they would rather not have this bill. . . . [W]hile I have no interest in preventing them from working out any equitable arrangements I do get the feeling they would rather not have the legislation.

Mr. Seiberling: They certainly don't sound very enthusiastic.

H.R. 2477 was favorably reported out by the full committee on May 10, 1983, then went to the House Agriculture Committee, which approved it on June 1, 1983. It went to the House floor on June 7, where only Rep. Manuel Lujan, R-New Mexico, spoke against it.

Lujan . . . asked the House to put off action on the Sipsey expansion until the Forest Service completes its land-use study for the forest. "It would make sense to wait until the forest planning process is complete," he said.

Flippo complained that the Forest Service, in opposing his bill, offered "the same old tired testimony the Forest Service offered last year." He said he was "shocked and dismayed" by the agency's stance, but added "we're getting used to [it]."

The main reason members of Congress should support the wilderness expansion, he said, is "the citizens of Lawrence County . . . want it maintained as it is and they want to share it with future generations."[7]

After eight minutes of this debate, the bill passed the House of Representatives by voice vote. SWUFFL's Marshell Frost was not pleased: "Why should citizens from the upper crust and the ranks of the wealthy influential, highly-educated special interest and professional-type people be able to come in and make their demands and wishes become law? Should we allow them to use the Bankhead National Forest for a playground and pet projects? Why should proponents of the bill who live 60–100 miles from the area be allowed to decide the needs of local citizens without their knowledge?"[8]

During the spring, we had begun to pursue a new and, we felt, persuasive avenue of justification for the Sipsey enlargement.

The original and largest source of public drinking water for the Birmingham metro region had historically been the Cahaba River. By 1983, however, the Birmingham Water Works was tapping Lewis Smith Reservoir to help meet the growing city's demands, with about 23 percent, or 8 billion gallons, of the area's annual drinking water supply coming from a pipeline to the Sipsey Fork just below Smith Dam. Since a great part of the impoundment was comprised of the West Fork Sipsey and Brushy Creek, it made sense to protect the source of this growing municipal water supply through wilderness designation at its headwaters.

With leadership from former Birmingham mayor David Vann, who now served as chairman of the Water Works Board of the City of Birmingham, the board adopted a resolution supporting H.R. 2477. Vann wrote to Heflin, "[T]his bill will add protection to the long-run water supply of the 700,000 or more people who depend upon our Board for the supply of drinking water. . . . [B]ecause our other sources of water are used to the maximum, all water for the expansion of Birmingham, Jefferson and Shelby Counties during the next 30–40 years must come from an increased use of water from this source."[9]

The Jasper Utilities Board followed suit, noting that its "plant provides the drinking water supply for the cities of Jasper, Cordova, Oakman, Parrish, and for several rural authorities, including those of Curry and Townley. We estimate that we serve approximately 45,000 people in Walker County."[10]

Additional endorsements came from the Bryan-Burnwell Water Authority, serving some 330 households; the City of Sumiton Water Works Board, serving approximately 9,200 people; and the Sipsey Municipal Water Works Board, serving around 7,200 citizens. "Because the continued prosperity of the Smith Lake area . . . is so directly dependent upon its clear waters and attractive, healthy surroundings," Flippo's legislation also received the endorsement of the Smith Lake Civic Association and the Walker County Commission.[11]

We hoped that this justification for protecting the watershed might provide a way to move Denton and Heflin away from their single-minded preoccupation with Winston County loggers to something more in line with Flippo's approach. After all, it was hard to argue with protecting the water supply for three-quarters of a million people.

Denton continued to be unreachable, doggedly insisting on waiting for the Alabama forest plan, even though everyone knew now what it would probably contain and despite an announcement by the Forest Service's Washington office that delays in the forest planning process would push release of Alabama's plan into 1985. Denton told the press, "There is so much emotion on both sides of this issue. . . . I've never heard of a wilder group on both sides. I'd like to see more objectivity."[12]

"The area has already been studied to death," said Pete Kelly, Ronnie Flippo's assistant. "That study was supposed to be completed by the end of this year. Now they're saying it will be a couple of years later." Denton and his staff remained in denial. "'The state forestry service [sic] has told us that it will be ready six to eight months before that timetable,' said Teresa Miller, the Denton aide who is handling the wilderness proposal."[13]

Following a jurisdictional dispute between the Senate Energy Committee and the Senate Agriculture committee over assignment of wilderness bills, H.R. 2477 went to the Agriculture Committee, where Howell Heflin was a member. This put the Sipsey expansion permanently and squarely in Heflin's lap, and Flippo's people urged us to meet personally with him as soon as possible.

First, we had to ask ourselves if there were some "compromise" that we could hold out to Heflin, give him cover to move the wilderness expansion in the Senate. Knowing that anything other than the death of the legislation would produce adamant SWUFFL opposition, we wanted to limit any concession we might make. The key, we felt, was the issue of watershed protection. Surely Heflin, after seeing Flippo's determination to enlarge the Sipsey and the public health advantages flowing from it, would take some meaningful action. What should we propose?

All but about 4,000 acres of the expansion proposal lay in the watershed of the West Fork Sipsey River. That stream had been designated by Congress in 1975 as a study river for potential inclusion in the National Wild and Scenic River System. The Forest Service was then conducting its evaluation, with the likelihood that the West Fork would be proposed by the agency for designation—indeed, that the Sipsey would be the only Alabama stream to get a favorable recommendation from the federal government. If an area were to be compromised, it had to be the 4,000 acres across the divide in the upper Brushy Fork watershed.

After some soul-searching, the "ruling triumvirate" of the Wilderness Coalition—Mike Leonard, Jim Price, and I—decided that we would propose to Heflin the deletion of upper Brushy Fork from the expansion as a supportable alternative. Frank Toohey in Flippo's office said it was worth a try.

Heflin agreed to see us in Alabama during Congress's Fourth of July recess. Mike Leonard, Luke Slaton, and I arrived at a restaurant in Decatur chosen by Heflin and found him holding court at a rear table. When our turn came, we produced a special map I had prepared that highlighted the canyons and waterfalls in the expansion area, all feeding south to join the West Fork Sipsey. After reviewing the importance of watershed protection for the West Fork, both in terms of its ecological value and its consequent impact on water quality, we offered our proposal to delete the Brushy Fork acreage from the expansion.

"Why, hell," exclaimed Heflin, "you want the whole thing!"

"Look," he said, "when I get back to Washington, I'm going to propose some alternatives as an opening bid. That won't necessarily be what we end up with, but I'm going to try to find some basis for a compromise." And with that, we were dismissed.

We had to learn what Heflin had in mind—as did Ronnie Flippo—by reading about it in the newspaper. Shortly before Congress's summer recess, Heflin called to his office several Washington correspondents for north Alabama newspapers and said, "We are exploring with a lot of different people on alternatives—decreased acreage, some potential of having the [existing] wilderness as a centerpiece with some increased acreage and then having the rest of the Sipsey in a classification which is known as a national recreation [area]."[14]

If Heflin had been exploring this national recreation area proposal "with a lot of different people," he certainly hadn't included anyone with the Wilderness Coalition or even Congressman Ronnie Flippo. Pete Kelly in Flippo's office told the press that "he is not aware of any discussions between Heflin and Flippo or even between the two congressmen's staff members."[15]

Heflin gave the press a list of the benefits, as he saw them, of a national recreation area (NRA) over wilderness:

A wider range of outdoor recreation opportunities.

Because of the diversity of activities, such a classification will "achieve optimum public benefit" and could include day-use centers, overnight and vacation campgrounds, trails for horseback riding and biking and scenic roads.

Commercial services would be required to locate outside the national recreation area in order to preserve the natural environment, "which could lead to development of private enterprise in the form of motel and other related lodging and feeding facilities."

Hunting and trapping may be authorized in a national recreation area. Timber harvesting may also be authorized, but the standards are much more restrictive than those associated with the multiple-use concept in national forests.[16]

Then, Heflin made some gratuitous remarks that forever changed the tenor of the Sipsey expansion effort:

"[O]utside of primarily the media, who seem to be greatly enthralled and attracted by this issue, it's not all that big a deal with most folks in Alabama. . . . Primarily the most ardent Sipsey supporters are an elitist group."

Editorial writers in Alabama, who have written numerous pieces strongly encouraging Denton and Heflin to pass [Flippo's] bill, "belong to that elitist group that envision themselves as birdwatchers and backpackers," Heflin said.[17]

Knowing that Congressman Flippo would want an informed response to Heflin, I put aside my anger over the senator's insults and embarked on some quick research into the legal niceties of national recreation areas.

There is no national system, as such, of NRAs, no standards of protection or management such as exist with the National Wilderness Preservation System. Each NRA Congress has created has its own rules, which vary according to the whims of its sponsors. Most NRAs are very large and can accommodate the kind of intensive development that Heflin had described. Perhaps the best example is Glen Canyon National Recreation Area, which encompasses hundreds of miles of shoreline at Lake Powell on the Colorado River above the Grand Canyon. Had Heflin any knowledge of the Sipsey watershed at all, he would have known that NRA designation for such an extremely small area was environmentally inappropriate. Moreover, had he done his homework, he would have known that an NRA was proposed as an alternative to the original Sipsey Wilderness in 1972 and was opposed then by the U.S. Agriculture Department because "considerable expenditure of public funds would be required to develop and administer the area, since the area probably would not attract private investment capital."[18]

It was wilderness, and the highest and best use of it was to preserve it.

Frank Toohey sounded very hopeful when I called to discuss the proposal with him and noticeably disappointed when I explained why we couldn't take Heflin's bait:

"Frank, Mr. Flippo's bill already provides for the Northwest Road to stay open and for horseback riding to be allowed. If Heflin wants to try to get an appropriation to build new trails or a new campground on the edge of the wilderness, we'd support him. For that matter, if he would agree to the Sipsey expansion and designate the entire Bankhead Forest as a National Recreation Area so that it would include the upper part of Smith Lake, we'd probably go for that, too. As it is, though, the NRA proposal is just a way for Heflin to look like he's doing something while keeping the area open for logging."

"Well," Toohey replied, "remember that you've got to get him to move something in the Senate." I suppose I was more offended by Heflin's ploy than I was eager to encourage him, since the press release I wrote said, "We don't understand why Sen. Heflin feels a National Recreation Area is needed or why he would propose something that may be meaningless.... Conservationists will carefully study Sen. Heflin's proposal when it is formally presented and we'll work in good faith to address our concerns, if he is willing. But in the meantime, we strongly feel that Senator Heflin owes the people of Alabama a straightforward answer as to why this great natural treasure is not entitled to the lasting protection of Wilderness."[19]

When Congress recessed in August 1983, Heflin conducted a number of open houses, or town meetings, around the state, and when he went to Moulton, he faced an angry group of Lawrence Countians.

> Lawrence County residents Monday disputed a published remark by U.S. Sen. Howell Heflin, D-Ala., that "the most ardent Sipsey supporters are an elitist group."
>
> State Rep. Roger Dutton said his grandfather is buried in the proposed wilderness area. "Being socially elite, he drove a garbage truck for the city of Moulton."
>
> "If I'm socially elite or Charles Borden is socially elite, or the people in this room are socially elite, then I think we've got some kind of communication problem," Dutton said.
>
> "We [Lawrence County residents] know the forest," said Dr. Charles Borden, a Moulton dentist. "We know it's being destroyed."
>
> "We of this area have a chance to have something that's unique," Lawrence County Probate Judge Rip Proctor said in support of the Sipsey addition.
>
> Heflin said ... that he opposes "excessive" acreage added to the existing 12,000-acre wilderness area in Bankhead National Forest.
>
> "He said he is seeking a compromise "in which maybe nobody will be terribly happy, but maybe nobody will be terribly mad."[20]

The *Moulton Advertiser* reported that "Heflin said when he made the comment about elitist backpackers and birdwatchers that he spoke in jest, but he defended the accuracy of the statement by saying that what he said was that the supporters included a group of elitist backpackers and birdwatchers."[21]

From Moulton, Heflin moved on to Decatur, and a few days later, to Birmingham, where he continued his call for a "compromise" on the Sipsey enlargement. There, he embellished a claim he had made to the media a few days earlier, when he asserted that the expansion envisioned by Flippo's bill would make the wilderness "about half [as large as] the average county in Alabama."[22] In Birmingham, he said the wilderness expansion would make it "one-third the size of Jefferson County." Heflin also claimed, as he had in Moulton, that hunting would be prohibited in the wilderness area.

Rebecca Falkenberry, a Sierra Club activist, attended the Birmingham open house and wrote to Heflin afterward that she was "shocked and upset by your attitude. . . . Some of your statements indicated not only insensitivity to the issue but also a lack of knowledge. . . . You stated that wilderness areas would 'not allow for hunting, boating and fishing.' Wilderness areas might not allow four-wheel drive vehicles for hunters . . . but certainly those activities of hunting, boating and fishing can occur. . . . You seemed to be surprised that Birmingham's water supply would be affected by the protection of the headwaters of the Sipsey River. And that Birmingham residents should be interested or concerned!"[23]

The *Moulton Advertiser* weighed in with an editorial:

Senator Howell Heflin made it plain . . . that he is not sympathetic toward the Sipsey Wilderness Additions bill as it stands after passage by the House of Representatives. . . . He also exhibited a lack of knowledge about the bill when he obviously didn't even know that hunting was allowed in the wilderness.

In a town meeting in Moulton . . . he laughed off a question about why he didn't support the bill after the house passed it two years in a row. "Congressman Flippo doesn't represent Winston County," he answered.

No he doesn't. But . . . why Winston County has the right to veto the permanent protection of Lawrence County land is a mystery to us. Heflin keeps talking about jobs being lost if the expansion goes through. The land hasn't even been in production for about six years, but all of a sudden, Winston County loggers are going to be out of work if the expansion is okayed. That's a pretty weak argument.[24]

The *Birmingham News* also felt moved to opine:

Heflin has said that he considers adding 29,000 acres to the Sipsey Wilderness too much and that he fears designating the whole area a "wilderness" will make it too restrictive.

The senator's charge . . . is . . . off base. If the land, already owned by the federal government, is granted wilderness status, hunters, campers, fishermen and others would still be allowed to use it. The only thing that would be barred, in effect, is timbering and mining.

Heflin says that he is mulling over a compromise that would see only a small portion of the land designated as wilderness, with probably the majority of it instead termed a "recreation area." That might be politically expedient, but it would, in effect, kill the intent of the bill. Both timbering and mining are permitted in "recreation areas."

Sen. Heflin usually appears to be reasonable on most issues. In this matter, however, that has not been true. It would be difficult to justify his position based on any objective look at the facts involved. As for Denton, he seems to be content to use his position on the Forest Service [plan] as a means to escape any involvement in the matter at all.

What Sens. Denton and Heflin appear to be saying is that somehow from Washington they know more about what's in the best interests of Alabama as regards this matter than do all [the public officials and groups that have endorsed the expansion].

They should reconsider.[25]

The *Anniston Star* took a more ecological approach:

Last week Sen. Howell Heflin charged that those supporting the Sipsey Wilderness bill are "an elitist group that want to backpack or be bird watchers."

It would be hard to conceive of a more inaccurate or misguided statement. It reflects an ignorance of what wilderness is all about. . . .

Wilderness is a matter of long-term survival of the diverse, complex, interacting living system of this planet. Wilderness is related to human survival itself. . . .

Without an environmental balance, [the] elimination of diversity is a threat to human survival. Wilderness, if there is enough of it, provides the diversity and complexity needed to ensure the survival of life on this planet.

Not enough has been done to provide for wilderness in the United States. Too many compromises are made by politicians under pressure from short-term special interests.

We urge both senators to drop rhetoric about "elitist bird-watchers and backpackers," and to understand the ecological issues behind the wilderness concept. We are not talking about luxuries here. We are talking about the future of life.[26]

As for me, I deliberately stayed away from Heflin's town meetings, as I was already angry enough. I felt that I had to challenge him, however, on his assertion that the Sipsey expansion was "half the size of the average county in Alabama" and "one-third the size of Jefferson County." I wrote Heflin a polite letter, with copies to several north Alabama newspaper editors, "to provide you with accurate information on this point":

> As you know, all of the proposed Sipsey Additions lie in Lawrence County. . . . According to the . . . Lawrence County Tax Assessor's Office, there are 439,050 acres in Lawrence County. The 28,000 acres of proposed Wilderness additions amount to just 6% of Lawrence County's acreage.
> Also, I'm informed that there are 687,976 acres in Jefferson County, meaning that the Sipsey expansion equals about 4% of the land here.[27]

No doubt, this did nothing to ingratiate me with Heflin, but at least he never made the same baseless claims again.

❧

Congress returned in September, and after several weeks of silence from Heflin, the press began asking if he were going to introduce his national recreation area proposal in the Senate.

"The extremes on both sides didn't take to it," the Tuscumbia Democrat said. "They haven't indicated too much willingness to move even an inch on it."

"My proposal was an effort to float an idea . . . to provide an area for compromise. But both sides have indicated that they will not move an inch. . . . I had hoped to hear more suggestions," he said, "but instead I got attacked over what I had said."[28]

My first impulse was to write and tell him how many inches were in 4,000 acres, but I suppressed the notion, thinking he had been challenged enough lately. I was also unwilling at this stage to accept the idea that Heflin might simply be hopeless.

Meanwhile, I attempted once more to open communications with the Alabama Forestry Association, going to its office in Montgomery and meeting with the outgoing executive vice president, Hilton Watson; the chairman of AFA's board of directors, Bill Lee; Union Camp's Bill Chestnutt; and John McMillan, soon to replace Hilton Watson. These gentlemen were very frank with me, or so I thought, as I reported to Howell Heflin in a letter: "AFA leaders have told me on several occasions that their Sipsey position is directly related to yours. Because you have not supported Mr. Flippo's legislation, the Association believes that they are acting according to your wishes by opposing the expansion. Thus, they will never go behind you to negotiate an agreement with us unless you give them your specific authority to do so. . . . The AFA Board has specifically authorized its staff to accept whatever solution you feel is necessary to bring this dispute to an end."[29]

Heflin responded:

I now realize that there has been a breakdown in communications. . . .

I gather . . . that you feel that you are making substantial progress with key leaders of the Alabama Forestry Association. However, in my conversations with them, I do not detect the same optimism for a solution that you articulate. . . . If you can reach an agreement or make substantial progress with this group, then I believe we will have accomplished a substantial first step towards finding a solution. . . . If you can work out something with the officials of the Alabama Forestry Association, I will endeavor to contact other opponents towards finding a solution to this problem.[30]

Was I wrong to interpret his remarks as meaning that even if we reached some agreement with AFA, "other opponents" could conceivably shoot it down? Did he mean he was going to give SWUFFL the last word? I wrote him back:

[O]ur opponents do not speak with one voice. With Congressman Flippo's close guidance, we have full authority to represent Sipsey supporters . . . and feel that we have the right to expect the same from our opponents. As a lawyer and a judge, you are certainly aware that no dispute can ever be amicably settled until both sides are represented by someone with full authority to negotiate. . . . [T]here needs to be a clear understanding that if an agreement can be reached, it will not be subject to veto by everyone claiming to have an interest in the issue. . . . [I] sense that [AFA] would be willing to represent Sipsey opponents in negotiations if they have your specific authority to do so.[31]

Again Heflin replied. "You make statements concerning the position of the Alabama Forestry Association which had not been communicated to me by the Association. I plan to discuss the contents of your letter with them and, hopefully, become better informed on their position."[32]

After that—nothing, nothing at all. Congress was heading toward concluding its 1983 session, and chances of action on the Sipsey before the recess were nil. The press asked for an update: "Heflin press secretary Jerry Ray indicated [that] 'nothing's changed' on the bill's status. 'Right now the attention in this place has been on Lebanon and Grenada.'"[33]

The *Moulton Advertiser* published a year-end editorial titled "Stalling Should End."

> [Howell Heflin and Jeremiah Denton] have chosen to heed the unfounded allegations of economic ruin put forth by Winston County's timber producers.... [T]he two senators have yet to make any serious attempt to work out a solution.
>
> Senator Denton has not even shown supporters of the bill the courtesy of meeting with them face to face. He sits in his ivory tower in Washington and sends messages through his aides, preferring to wait until the bureaucrats in the pro-timber-producer Forest Service give him their opinions.
>
> While Senator Heflin has met with wilderness supporters on several occasions ... his only gesture toward a compromise has been the suggestion of giving wilderness protection to a very small bit of additional acreage and categorizing the rest as a "national recreation area," which would offer no protection at all.
>
> [T]he economic question has been satisfactorily dealt with. The question that has not been answered is why the greed of the logging lobby continues to take precedence with the senators over the wishes of the thousands of Alabamians who want to preserve this beautiful piece of their state.[34]

CHAPTER EIGHTEEN

Nobody Loves You but Your Dog

THE ONLY POSITIVE THING THAT could be said about the failure of Flippo's legislation to move in the Senate during 1983 was that it did not have to be reintroduced and passed by the House again in 1984, as each Congress ran for two years. The 98th would not conclude until late fall of 1984, so there was still hope that something might succeed in time.

Clearly, however, Howell Heflin was digging in. "The extremes on both sides don't want to give much," he again told the press, adding that this is "one of those things that will probably take a while to work out.... Right now, we're letting it simmer a bit and letting the emotions settle down.... Maybe a little waiting period might calm down a few emotions."

Ronnie Flippo's public statements reflected an attempt to mollify the senator. "There's no animosity here.... I believe what I'm trying to do is right and he believes what he's doing is right. I have [a] very high regard for Senator Heflin."[1]

Nevertheless, the burden remained with the Wilderness Coalition to find a solution. Having effectively killed the national recreation area idea to Heflin's public embarrassment, our challenge was to reestablish communication and urge him in the right direction.

In December 1983 I asked Mike Leonard if he would undertake a "peace mission" to both senators. This proved to be a better idea than perhaps even Mike or I realized at the time, as it gave him his first real opportunity to enter a hostile environment and bring to bear his natural ability for influencing public officials. In describing Mike, perhaps the best analogy I can make is to that of the successful corporate rainmaker, who is completely at ease in the world of country clubs, golf cleats, and boardrooms, not because he has determined to make himself that way, but because such things are as inherently attractive to him as to those he is seeking to schmooze. Mike admires politicians and the things they do, and he makes that admiration clear, as well as a deference for their official positions. He understands what it takes for politicians to succeed. They recognize this trait in Mike and, consequently, trust him. There was a time during the Sipsey expansion fight that Howell Heflin refused to discuss the matter with anyone from the Alabama environmental community except Mike Leonard.

All that notwithstanding, Mike's talents did not produce any immediate results. By spring, he had met twice with Heflin and once with Denton and had succeeded only in nudging them a little way from their preoccupation with having been so roundly criticized by the environmental community and the state's news media. If there had been any progress with Heflin in particular, it was threatened by Charles Borden's decision to run against him in the Democratic primary for the U.S. Senate.

"All public officials have a responsibility to have some rational explanation for their positions," says Charles. "After we had given him all our reasons to protect the forest, and we had been reasonable, and had demonstrated all the public support that we had, he didn't care, he was opposing it because of some sort of personal distaste for the issue. I felt that someone had a responsibility, that *I* had a responsibility, to voice opposition to him on this issue particularly. I felt that I had to attack him in the marketplace of ideas."

Borden dogged Heflin throughout the primary campaign, going to "dozens" of meetings. "Where he spoke, I spoke. If I could find a meeting where he was going to be speaking, I went as a candidate and spoke. I took my candidacy for the Senate as a great responsibility, and I took it seriously."

In early 1984 the Forest Service introduced a couple of new dynamics into the situation. The first was its formal recommendation of the West Fork Sipsey for inclusion in the National Wild and Scenic Rivers System. The second was a policy being developed at the regional forester's level that effected something of a reversal in the Reagan administration's previous stance toward RARE II areas. Publicly announced in June 1984, the new guideline concluded "that all of the proposed RARE II wilderness areas and 55 percent of the further planning areas [in the southern region] would require wilderness designation to meet [agency] targets," raising the possibility of increased wilderness acreage in the still-pending Alabama National Forest Management Plan.[2]

It was abundantly clear now that Howell Heflin would never support either Ronnie Flippo's original 28,000-acre enlargement or our proposed deletion of the 4,000 acres in upper Brushy Creek. The injection of the wild and scenic river recommendation into the mix posed both problems and promise. The initial problem was that the draft study recommended the West Fork Sipsey River itself for designation but excluded all the upper tributaries that were contained in Flippo's legislation, even though those streams were a part of the formal study area.[3] While we enthusiastically supported inclusion of the river in the national system, we feared that the Forest Service's limited recommendation might be employed as a substitute for a meaningful wilderness expansion, as had been the case with the national recreation area proposal. Senator Heflin hinted as much when he told the press, "I am optimistic that the designation of the West Fork of the Sipsey within the Wild and Scenic Act can lead to an agreement which will bring about a reasonable expanded Sipsey Wilderness."[4]

The promise of the wild and scenic river proposal lay in the possibility that the corridor could be expanded at the legislative stage beyond that recommended by the Forest Service, so as to include the canyons of any upper tributaries that Heflin or Denton might want to cut from the wilderness expansion. This definitely had to be a fallback position for us. While statutory designation of the canyons themselves was of course desirable, their environmental health depended on the integrity of the divides above them, which wilderness designation would provide. We had to be very careful about how we presented the notion of expanding the wild and scenic river corridor, again fearing that it might be used as a politically expedient way to prevent a significant wilderness enlargement.

After first getting the Birmingham Audubon Society's approval, I called a Wilderness Coalition meeting to float an idea I wanted to pursue: first, along with upper Brushy Creek, I proposed that we delete the RARE II area known as Montgomery Creek, which encompassed some 6,000 acres of Borden Creek's upper tributary. This would produce a 10,000-acre reduction from Flippo's original legislation, but it would leave intact all the watersheds of Tedford, Thompson, Braziel, and Hagood Creeks, which, in fact, were the wildest of the Sipsey's upper reaches. This would produce a wilderness expansion of just under 18,000 acres. Second, we would cautiously and privately seek Ronnie Flippo's agreement to sponsor a wild and scenic bill that would include Montgomery Creek in its corridor.

This idea naturally caught the Wilderness Coalition by surprise and produced some resistance, particularly on the part of one James L. Taylor, a newcomer to the Sipsey fight who now requires an introduction, as he became a staunch ally in the effort.

Jim Taylor is a native of the east Texas town of Madisonville. His first career was as a Marine Corps officer, retiring with the rank of major. After earning a Ph.D. from Columbia University, he accepted a teaching position in the School of Commerce and Business Administration at the University of Alabama. He was introduced to the Sipsey Wilderness by the university's Doug Phillips and became so attached to it—he calls it "one of the greatest places in the world"—that he converted a business course called "Resources of the United States" into an environmental education course so he could take students to the Sipsey. By early 1984 Jim Taylor had been elected chairman of the Sierra Club's Tuscaloosa Group, a position that led to his participation in the Wilderness Coalition. His involvement came at an opportune time, as Jim Price was being increasingly consumed by his regional duties as the Sierra Club's southeastern representative and was finding it difficult to remain closely involved with the Alabama Wilderness Coalition. Jim Taylor became the first Sierra Club volunteer leader since Wally Retan to take an active role in the wilderness effort.

Taylor was not happy with my idea of unilaterally dropping Montgomery Creek. He had studied the art of negotiation and knew that concessions should be made only grudgingly. He felt that "we should have retreated a little slower. Who are we negotiating with?" he asked. "We're compromising with ourselves!"

"Very true," I said. "That's exactly our problem. We don't have anyone to negotiate with. Heflin won't do anything unless we 'compromise' with the timber industry. The timber industry won't negotiate because they know that Heflin won't do anything. As long as Heflin can paint us as extremists unwilling to reach some accommodation, he'll begin to erode our support. Ronnie Flippo is a politician, too. He's looking for a way to succeed with this. We've simply got to come up with something new for him to work with."

In the end, Jim and the others in the coalition accepted the dilemma we faced, and Mike Leonard was asked to present the new 18,000-acre proposal to both senators. We resolved to stay out of the press and let the congressional delegation chew this new idea over in private.

The second Forest Service initiative, the conclusion by the regional forester's office that the agency should consider 55 percent of the RARE II "further planning" areas for wilderness designation, translated into 16,000 additional acres for Alabama, if accepted by planners in the state office. On March 30 I traveled to Montgomery to meet with Alabama forest supervisor Joe Brown, both to promote our 18,000-acre "compromise" and to get his reading on the regional forester's new policy.

Brown said they were now looking at 15,000 more acres in Alabama, but they felt only 10,000 acres in the Bankhead (the watersheds of Braziel and Hagood Creeks) were appropriate, and that the additional 5,000 acres should be located elsewhere in the state in order to spread around the impact on timber resources. The new area they had in mind was Dugger Mountain in the northern Talladega Forest. I told him that just as we had been unwilling to trade the Cheaha Wilderness for Dugger Mountain, we did not want to substitute it for the Sipsey watershed. "Now that we've drawn so much closer in our positions," I said, "I hope you'll see the benefit of helping us bring this thing to a conclusion without quibbling over 2,000 or 3,000 acres." Brown was noncommittal, but he authorized me to summarize our conversation in a letter to the congressional delegation, which I did.

Over the course of the two years that the 98th Congress was in session, either Mike Leonard or I, or both of us, met five times with the leadership of the Alabama Forestry Association in an effort to reach some consensus. During our final meetings, the association indicated that they would acquiesce in the Forest Service's new notion of a 10,000-acre Sipsey expansion, but they made it clear that they did not want to be publicly perceived as actually supporting *any* enlargement of the wilderness.

One more meeting with Joe Brown produced what we hoped might finally induce action in the Senate. Although Brown was unwilling to budge on his 10,000-acre projection for the Bankhead, he did say, for the first time, that the Forest Service wanted to get the wilderness issue handled by Congress before it released its Alabama forest plan. This was a significant reversal of position. We urged Brown to communicate this to Denton and Heflin. If he did so, the senators weren't buying. Denton continued to state his intention to wait for the forest plan, and Howell Heflin began to express the same sentiment.

Nor did any of Mike Leonard's peace overtures do much good. After an initially encouraging response from both senators' staffs to our new proposal and the shifts in Forest Service thinking, Heflin got his back up again because of Charles Borden's campaign against him for the Democratic nomination. Mike tried his best to placate Heflin, telling him, "Senator, none of us has any control over Charles Borden. We couldn't even get him to wear a tie to testify in the House!" In point of fact, Borden's Senate run was never anything more than a token effort, and Heflin trounced him 399,817 votes to 47,463.[5]

Instead of brushing Borden off as the minor candidate he was, Heflin saw his victory in the primary as a vindication. Heflin told Birmingham congressman Ben Erdreich that he was "personally offended" that Borden ran against him, apparently giving this more importance even than Ronnie Flippo's overtures to him. "I really want this bill. . . . I'm willing to be reasonable," Flippo told the press, to no avail.[6]

Today, Borden says he was unaware that Heflin used him as a justification for opposing the Sipsey bill. "I had not heard that before. That's a level of pettiness that I would not have expected from him. I would think that was just a convenient excuse, rather than a valid reason."

In the waning days of the 98th Congress, eight southeastern states successfully passed national forest wilderness legislation: Texas, 34,400 acres; Arkansas, 91,000 acres; Mississippi, 5,500 acres; Tennessee, 24,942 acres; Georgia, 14,529 acres; Florida, 49,150 acres; North Carolina 68,750 acres; and Virginia, 64,584 acres, a total of 352,855 acres. Nationally, the figure was 8.3 million acres. The Sipsey expansion, now dwindled to 18,000 acres, died in the Senate Agriculture Committee because of its purported threat to the timber industry.

Ronnie Flippo, according to a memorandum circulated to southeastern activists by the Wilderness Society, "is incensed by Heflin's tactics and his opposition. Privately, Flippo has threatened to introduce a bill next Congress that would protect all of the state's RARE II areas, essentially locking out any form of development in the entire inventory. While this is likely an idle threat, . . . failing to get legislation such as this for one's own district two Congresses in a row can be very embarrassing for a politician."[7]

What Ronnie Flippo actually decided to do caught everyone by surprise.

"John," said a somber Frank Toohey in a telephone call, "Mr. Flippo has decided that he will not reintroduce his bill next session. He's tired of the senators doing nothing. You guys have got to get one of them to move on this, and then we'll see what we need to do in the House."

Flippo expanded on this to the press:

[Flippo is] plainly frustrated. "I see nothing to be gained by me again demonstrating that I can pass [a Sipsey bill] in the House. . . . We have passed wilderness bills for

30 or 40 states in the past two years involving millions and millions of acres of land. ... And there is no part of the Sipsey Wilderness that's not well understood. I guess it's unrealistic to expect the forestry folks and the environmental folks to agree with each other a hundred percent, but both of them have stated their cases. Now it's time for those on the Senate side to decide what they want to do with it, and if they decide what they want to do, I'll be here."[8]

We had reached the low point in the Sipsey expansion campaign. The message from Flippo was clear: We were teetering on the brink of losing our House sponsor altogether. We were on our own. If we expected an enlargement of the Sipsey Wilderness, then we had to do whatever was necessary to get the senators' support. Heflin wanted more "compromise," telling the press that the two sides are "still a good ways apart. . . . [I]t appears to be stalemated."[9]

Our dream of statutory wilderness protection for the watershed of the West Fork Sipsey—or even a meaningful part of it—was dead. Worse, with no Sipsey expansion legislation passed by the House, the Forest Service would be free to open up most of the watershed for timber production once its Alabama forest plan was formally adopted. Within the ranks of the Wilderness Coalition, there was some talk that, like Charles Borden, I was too caustic, too confrontational, and should be replaced as leader of the Sipsey effort.

"Poor John," said Mike Leonard. "Nobody loves you but your dog."

CHAPTER NINETEEN

Stay Out of My Business

WHATEVER LESSONS 1984 HELD FOR US, clearly, staying out of the press and trying to be diplomatic toward the Senate delegation had been a failed strategy. As one reporter put it, "Alabama members of Congress . . . seem to have lost interest in the Sipsey issue."[1]

The *Anniston Star* reported that

[Senator] Denton's press secretary, Irma Engelkes, says Denton feels it's up to Heflin to push a bill through the Senate Agriculture Committee, since Denton is not a member. Heflin aide Steve Raybe [sic] is non-committal. He says the Democratic senator will wait to see the Forest Service [land management plan].

"We're not going to stand [for] any more [wilderness]," says Marshall [sic] Frost, owner of a $2 million a year lumberyard in Haleyville. "There's 31,000 acres up there that needs to be timbered."[2]

And Frost told the *Birmingham News*, "We've been successful at getting them [the senators] to do nothing, and we hope that they will keep doing that and let it die."[3]

Mike Leonard decided to resign as "negotiator" for the Wilderness Coalition. He wrote to Heflin's assistant, Steve Raby, "I don't feel like the negotiations that have already gone on have done much good insofar as the timber industry's position is concerned. . . . [H]ow can someone 'negotiate' when there is no one who seems to have the authority to speak for the other side of an issue. Or, more specifically, how can I negotiate in good faith with the Alabama Forestry Association when . . . [SWUFFL] feels free to . . . ignore what those who are willing to negotiate have accomplished?"[4]

Raby, in turn, chose to deflect responsibility away from the Senate delegation. "The problem is with the people in the state. Until they get together, there's not a lot we can do," he told the press.[5] A *Decatur Daily* editorial responded: "[L]oggers and conservationists are so implacably opposed that asking them to independently compromise is much like asking the foxes and chickens to work out a security agreement before the farmer will build a fence."[6]

We were compelled to "go public" once more, and our first step was to announce the 18,000-acre compromise that we had privately communicated to the senators during 1984. Our press release read:

> Our new proposal is a good faith effort to give our Senators a solid middle ground on this issue. It represents a reduction [10,087 acres] of more than one-third of the original proposal, a reduction almost as large as the present Sipsey Wilderness Area itself. On the other hand, it pares down the expansion to the wildest, most rugged areas, containing the largest concentration of old-growth hardwoods.... While it saddens us to give up an important watershed like the [Montgomery Creek] area, we felt that we should concentrate on saving these last old hardwoods from clearcutting and conversion to pine.[7]

The result was renewed press attention on a statewide basis. The *Birmingham Post-Herald*'s Washington correspondent, John Brinkley, wrote a perceptive analysis of the situation entitled, "Conservationists Losing Fight over Sipsey":

> The fight to annex a substantial portion of the Bankhead National Forest to the Sipsey Wilderness looks, at this point, like a losing one. The logging interests, who are eager to have the disputed forest acreage set aside for their chain saws are clearly in the driver's seat now.... Heflin, meanwhile, continues with the same line he has been using for some time. "My position of flexibility and compromise remains consistent." ... The Judge says he is waiting for something to come of the negotiations between the loggers and the conservationists before he does anything. That sounds reasonable, except for one thing: there are no negotiations.... And why haven't there been any meetings? Because the loggers know that if the present state of affairs continues, they will come out on top and therefore have no reason to negotiate.... Randolph and his cohorts are starting to panic. They see that land slipping away for good and nobody doing anything about it. If Heflin and Denton are really waiting for the opposing forces to reach a compromise, they have an awfully long wait ahead of them.[8]

While we weren't exactly "starting to panic," things were definitely grim and we needed to do something to regain the initiative. In the absence of pending legislation, we turned our attention to the Alabama National Forest Land Management Plan, which was finally released in draft form for public comment on April 1, 1985. As expected, the draft recommended the roughly 10,000-acre Sipsey expansion Joe Brown had described to me in our March 1984 meeting. The precise acreage figure was 9,793 acres, encompassing the watersheds of Hagood and Braziel Creeks and part of Borden Creek. The difference between this and our 18,000-acre compromise was the watershed of Thompson Creek.

Unless challenged through an administrative appeal, the plan was due to be implemented on October 1, and the remaining 18,500 acres that had been included in Flippo's initial legislation, including Thompson Creek, would be released for commercial timber production. Already, the shut-down sawmill in Grayson had been reopened by a new owner, ready to begin logging the Bankhead as soon as the new plan was implemented. "That's why we opened the mill and created jobs for 30 or 40 people," the new general manager told the press.[9] (In point of fact, though, the Forest Service had already made accommodation for this and other timber mills seeking to log the Bankhead. According to the draft forest plan, even if our 18,000-acre compromise were implemented, the Forest Service still planned harvesting over the ten-year life of the plan that averaged a 63 percent increase over current levels. By this time, however, the myth of harm to the timber industry had morphed into fact, and no one paid any attention.)

We had been working with national conservation groups on the possibility of an appeal of the forest plan in the event no legislation was introduced in Congress. On April 25, 1985, the Sierra Club Legal Defense Fund and the Wilderness Society issued a joint press release announcing their intention to file an administrative appeal and, ultimately, a lawsuit challenging the adequacy of the Alabama forest plan.

Ron Tipton, the Southeast Regional Representative of The Wilderness Society . . . said that, "Our initial review of the Draft Plan reveals several inadequacies relating to the Wilderness issue alone that might support a legal challenge. For example, none of the alternatives considered by the Forest Service contain the same lands as those in Congressman Flippo's legislation. There is no site-specific justification for excluding certain areas from Mr. Flippo's proposal and including others, nor is there adequate consideration given to protecting other possible Wilderness Areas in the state. . . . [The] paltry 9,793-acre proposal for the Sipsey makes the Alabama Draft Plan stand out like a sore thumb."

[Rick Middleton, staff attorney for the Sierra Club Legal Defense Fund said that] "if we undertake to challenge the Alabama Plan, it will be with a view toward reforming National Forest policy [in Alabama].... [W]e would hope to work a major reform in the Forest Service's policy of skewing National Forest management towards timber production at the expense of the environment and the federal taxpayer."[10]

SWUFFL's Marshell Frost responded, "If they file suit against the Forest Service, we're going to file suit against them for a billion dollars for keeping all that timber protected since 1977."[11]

Most of the state's newspapers—including the *Florence Times-Daily*, the *Birmingham News*, the *Moulton Advertiser*, the *Decatur Daily*, the *Huntsville Times*, the *Anniston Star*, the *Montgomery Advertiser*, the *Birmingham Post-Herald*, the *Alabama Journal*, and the *Jasper Daily Mountain Eagle*—published persuasive editorials urging Denton and Heflin to halt the impasse and support the Wilderness Coalition's 18,000-acre compromise.

And then, the renewed public outcry produced the first positive news for the Sipsey expansion in years, and it came from a startling source: Republican senator Jeremiah Denton. In the spring Denton began circulating a letter that read in part, "It is my intention to initiate legislation during this session that will offer a solution to the Sipsey question.... [I] hope you will be reassured in knowing that I plan to support legislation that will resolve the issue once and for all."[12]

"But," reported the *Birmingham Post-Herald*, "he is not talking—even to Flippo—about what that legislation might be. One would think he might communicate with Flippo about this mysterious bill, but Flippo's people say they haven't heard a word from him"[13]

Denton offered this explanation to the press: "I have switched to where I am more concerned about taking care of the environment than I was when I first came up here." The *Birmingham News* reported that "Denton said he's been 'taken aback' by unfavorable reaction to the Forest Service study he wanted to support. 'I thought the study would be a sober, responsible analysis of it,' said Denton. 'But the study doesn't seem to go far enough to suit the newspapers.... I'll get with Heflin right away on it and see if we have a joint position.' . . . 'I look forward to looking at any ideas Sen. Denton might have,' Heflin said. . . . Denton hadn't talked to him yet about the issue, he said."[14]

Actually, Denton had no specific proposal in mind. Instead, he instructed his in-state director, Richard Lee, to open communications with the various factions—the Wilderness Coalition, the Forest Service, the Alabama Forestry Association, and SWUFFL—to see if a middle ground could be identified. After first meeting independently with each group, Lee called for all the parties to gather in Denton's Montgomery office on July 11, 1985. Heflin had been uninvolved up to this point, but Lee invited him to send a staffer to participate in the meeting.

In preparation, we met first with Alabama forest supervisor Joe Brown. Joe had a new proposal to make, this one for about 12,350 acres, which added the southeastern third of the Thompson Creek RARE II area to the 10,000 acres recommended in the draft forest plan. Having been encouraged by Flippo's staff to reach an agreement if we could, I decided at last to raise the issue of expanding the wild and scenic river corridor. Reminding Brown that the "official" study area had specifically included the Sipsey's upper tributaries, I told him that his new proposal might be workable if the Forest Service would also recommend all the major tributaries for inclusion in the wild and scenic river corridor. The look of dawning optimism that then came to Brown's face was one of the more memorable events of the long Sipsey campaign.

And so, on July 11, our delegation—Jim Taylor for the Sierra Club, Mike Leonard and Ed Passerini for the Alabama Conservancy, and I—traveled to Montgomery for Richard Lee's set-to. Boyd Kelly of the Alabama Forestry Association appeared, and Heflin sent an in-state staffer. But a new face was there to represent SWUFFL: Charlie Mitchell, freshly resigned from Howell Heflin's staff and the man who Ronnie Flippo believes was most responsible for persuading Heflin to oppose the Sipsey expansion. If nothing else, Mitchell's appearance now as the official SWUFFL spokesman was revealing of the mind-set we had been up against.

Joe Brown presented his new proposal, which became known as the Montgomery plan: the 12,350-acre wilderness expansion that he had privately described to us, plus wild and scenic river designation for Thompson, Tedford, Flannagin, and Montgomery-Borden Creeks, as well as for the river itself, an additional protected area of some 4,400 acres. Moreover, Brown said, the Forest Service would adopt a semiprimitive administrative designation for all the RARE II lands that had been included in Flippo's original legislation. "Semiprimitive" defies easy definition, but it was characterized by smaller clear-cuts, a slower pace of harvesting, and other limits on timber operations.[15]

"Gentlemen, this looks good to me" said Richard Lee, "what do you say?"

Based on our private meeting with Brown, this proposal was what we had expected him to make, and we had already decided, despite sincere misgivings and some disagreement among ourselves, to accept it. Again, our focus had to be to induce one of the senators to take action and then let Ronnie Flippo take over. "We'll agree to that," we said.

"Gentlemen," said Lee, "we have an agreement. I'll submit it to the senators for their action."

The timber industry folks sat there with their mouths open. "I don't have authority to accept this," said the AFA's Boyd Kelly. "We'll have to submit it to our board of directors."

"Well," replied Lee, "this is what I'm giving to Senator Denton." Privately, Lee told us later that he appreciated our willingness to work things out and complained that the timber industry had been obstructionist and uncooperative. Publicly, he told the press, "'I've grown weary' of working further on a total agreement between the groups. Even those who did not voice support for the final plan . . . agreed the time for a decision had come, Lee said. 'After we finally came down to the bottom line, we felt it was time to make a decision if it was going to be handled in the Senate this year,' Lee said. 'The senators will make the final decision.'"[16] "We're going to do what we've got to do, [Lee said]. . . . The ax is going to fall."[17]

SWUFFL's Marshell Frost said, "There has been no compromise. We've all agreed on that. . . . We'll see it die in the Senate again before we'll go along with that."[18] The AFA's Boyd Kelly agreed. "I don't think I've got one chance in a million of getting that approved," he said.[19] "I've talked with everybody who gives a damn, [said Kelly], and this is a pipe dream for me to sell this."[20]

Nevertheless, Jeremiah Denton continued to promise action:

U.S. Senator Jeremiah Denton said . . . that he expects to introduce a bill in September to expand Alabama's Sipsey Wilderness Area. Environmentalists and timber interests are moving close to a settlement on the amount of acreage at about 12,000, said Denton. "At some point . . . I will say, 'Okay, this is the ballpark, I think it ought to be settled,'" said Denton. "I'm hearing a whole lot from people who make money out of it. I wanted to make darn sure I'm hearing from the people who just aesthetically enjoy it. . . . I don't think we should diddle around with this forever."[21]

The timber industry complained that they had not been given a fair hearing and requested another meeting to submit a proposal they wanted to make. So on August 7, 1985, the groups gathered again, this time in Denton's Birmingham office, and this time with the glare of television lights as well as the presence of the print media. SWUFFL brought a large delegation, while the Wilderness Coalition was represented only by Mike Leonard, Jim Taylor, and me.

"Our position had always been that there should be no more wilderness whatsoever," said Charlie Mitchell.[22] Nevertheless, to provide an "end to the controversy," SWUFFL and the Alabama Forestry Association distributed a new proposal: enlarge the Sipsey by the some 7,000 acres that had been recommended by the Forest Service at the conclusion of RARE II, expand the wild and scenic river corridor to include the upper tributaries, as Joe Brown had proposed at the last meeting, and establish a new wilderness area of some 8,400 acres at Dugger Mountain in the Talladega National Forest. The timber representatives wanted the western half of the Thompson Creek area open for general logging, with only the eastern half designated semiprimitive; otherwise, the proposal adopted the Forest Service's semiprimitive recommendations for the rest of the Bankhead RARE II areas.

The rank-and-file SWUFFLites in attendance seemed to be in denial concerning their own proposal. "I wouldn't care if there wasn't a dab more wilderness put in," said Marshell Frost. "We don't want people coming up there taking our timber land away from us."[23] Another man said that "the only people who want wilderness are rich bird watchers," to which retired U.S. Marine Corps major Jim Taylor replied, "If you think I'm rich, you haven't seen my tax returns, and I don't know a buzzard from a hummingbird as far as bird watching is concerned."

Richard Lee put an end to all this by announcing a new plan, one that he and Joe Brown had devised. It was a bastardization of the Montgomery plan, cutting acreage out of northern parts of the Thompson, Hagood, and Braziel Creek RARE II areas so as to pare the Sipsey expansion down to roughly 10,300 acres. The enlarged wild and scenic river corridor and the administrative semiprimitive areas remained as previously proposed, but now Lee and Brown were also adopting the timber industry's idea of an 8,400-acre Dugger Mountain Wilderness Area.

"This is it," said Lee. "This is going to the senators."

"I'm glad it's over," Lee told the press.[24] Joe Brown "said the plan proposed by Denton's staff was the most likely to be passed by Congress. 'I won't be relieved until it's finally over. I've been working on it for four years,' Brown said."

Unfortunately, the others weren't buying it. "We feel like we've bent over backwards with our proposal here," said Charlie Mitchell. "'I am doubtful that the Alabama Sierra Club or The Alabama Conservancy will support the plan,' said Jim Taylor. . . . '[We] bent over backwards to reach the Montgomery Plan . . . and we weren't very happy with that.'"

Even though the new proposal was truly awful in terms of the hatchet job it performed on the Bankhead RARE II areas, I tried to put a positive face on the situation in hopes of stimulating some action by Denton: "John Randolph of the Birmingham Audubon Society said the Denton staff compromise went 'much further towards [meeting] our concerns than the timber-industry proposal.' But, he said, if Heflin and Denton 'both support it [the timber industry proposal], then obviously, the conservation community would have to look at it pretty carefully.'"

Interestingly, SWUFFL's Bill Bustin was apparently thinking along the same lines: "Toward the end of the 2½-hour meeting . . . Bustin looked at Denton's state director Richard Lee and said, 'I think it's time you took over and did what you need to do.'"

"I think [this plan] will fly," replied Lee.

The press sought reaction from the congressional delegation, now including Congressman Bill Nichols, since Dugger Mountain was in Nichols's district and would require his support to be enacted:

"We don't want to get in the position . . . where we're supporting Dugger Mountain at the expense of what Ronnie Flippo wants for his district," [Nichols's aide] Winston Lett said. "If Mr. Flippo is for it, then we'll have no objection to Dugger Mountain being added."

Meanwhile, Flippo's position on the latest Sipsey compromise is the same as it has been in months past. "He's going to wait and see what the Senate does," said press secretary Mike Adcock. "He's not really going to comment on it until that point."

Heflin, meanwhile, issued a non-committal statement. "I haven't seen Sen. Denton's proposal," he said. "I don't know what he's proposed. I'll be interested in seeing it and studying it."[25]

Clearly, the initiative rested with Jeremiah Denton. To encourage him, we embarked on a round of letters to the editor and guest TV editorials praising Denton. My letter read:

Alabama conservationists are deeply grateful to Sen. Jeremiah Denton for his leadership in attempting to achieve a fair compromise. . . . Whatever one may feel about the specific proposals that have been presented, several important things have been accomplished.

First is that, at long last, there apparently will be Sipsey legislation moving in the Senate this fall. . . . Sen. Denton has undertaken to brave the wrath of the state's timber industry to get a Senate proposal moving, and he deserves all the credit for his statesmanship.

Second, the negotiating process has forced the state's timber industry—after nearly four years of surly intransigence—to spell out a specific Sipsey proposal that it will support. As paltry as the proposal may be . . . it nevertheless provides an alternative for legislation in the event either of our senators feels compelled to support the timber industry's position. The important thing is that the Senate pass some legislation so that, at the very least, a compromise can be achieved in a House-Senate conference.

Finally, and most importantly, the responsibility for resolving the Sipsey issue at last has shifted to where it belongs—in the hands of Sen. Jeremiah Denton, Sen. Howell Heflin and Rep. Ronnie Flippo, who were elected to make decisions like this.[26]

In retrospect, though, I should have known Heflin would never let himself be perceived as such a complete tool of the timber industry as to actually support the plan it had presented to Denton. Mike Leonard puts it this way: "The funny thing about Heflin was that he didn't want anything to happen, number one, but frankly if anything did happen, he wanted credit for it. So here was Denton doing something, and I think Heflin was afraid Denton would get all the credit for it and Heflin wouldn't."[27]

Heflin made himself clear to Denton very shortly. In a classic encounter in a Senate Office Building elevator, Heflin "jumped all over" Denton, saying, in so many words—Why are you doing this? There aren't 500 people in the whole state of Alabama that care about the Sipsey Wilderness. You're just doing this for the publicity since you have to run for reelection next year. This legislation is in my committee. Stay out of my business.

That worked. Denton did not sponsor legislation as he had promised. And the AFA's John McMillan told Mike Leonard in September that Heflin's problem with Charles Borden was going to prevent his agreeing to any legislation again that year. Once again, Heflin's determination to stall the Sipsey expansion held sway.

Mike Leonard happened to be in Richard Lee's Birmingham office shortly after the word came down from Washington that Denton wasn't going to introduce legislation after all. Lee had all the various press clippings that covered his negotiating sessions hung on the wall near his desk. Pointing to the clippings, he told Mike, "I'd like to shove these up Denton's ass!"

So, once again, the burden was back on the Wilderness Coalition to try to make something happen. Mike Leonard scheduled a meeting with John McMillan and Charlie Mitchell at a restaurant in Prattville, and there they devised a proposal that came to be known as the gentlemen's agreement. First, it called for a Sipsey expansion of 10,890 acres encompassing the entire watersheds of Hagood and Braziel Creeks and part of Borden Creek. Next, wild and scenic river designation would be extended to all the upper tributaries (Tedford, Thompson, Mattox, Braziel, Hagood, Flannagin, Borden, Montgomery, and Horse Creeks) which, along with the river itself, encompassed combined acreage outside the wilderness of some 5,500 acres. Third, the Forest Service would be asked to designate the excluded RARE II areas in the West Fork Sipsey watershed as semiprimitive, another 18,000 acres. And finally (since Winston Lett in Congressman Bill Nichols's office continued to say it was acceptable) there would be an 8,400-acre wilderness area established at Dugger Mountain in the Talladega Forest.[28]

While we were sorely disappointed at the limited wilderness expansion, our coalition groups had to admit that this proposal did achieve, through a mix of statutory and administrative designations, a degree of protection for the West Fork Sipsey watershed that was far superior to anything that had existed before. The semiprimitive designations at least held open the possibility of a future wilderness expansion when and if the makeup of the congressional delegation ever made a new campaign feasible. The attention given Dugger Mountain by the Forest Service from the beginning of RARE II to the present had induced more folks to become familiar with it, with the result that there was growing grass-roots support for a wilderness area there. Most importantly, the gentlemen's agreement would produce at last the "compromise" with the timber industry that Howell Heflin had been insisting upon. We decided to accept it.

Mike Leonard met with Heflin to tell him about it. Heflin was "just stunned" that an agreement had been reached, says Mike. "He had been telling me over and over to try to see if I could work out some compromise, thinking it would be impossible. Now here we were with an actual agreement."

To underscore the new consensus, we decided to send a joint delegation to Washington to present the gentlemen's agreement to the senators and congressmen. Thus in November 1985 Joe Brown of the U.S. Forest Service, John McMillan of the Alabama Forestry Association, Jim Taylor representing now both the Sierra Club and the Alabama Conservancy, and I, representing the Birmingham Audubon Society, arrived in the nation's capital with some hope of getting the Sipsey issue resolved at last.

John McMillan wanted to meet with Heflin in advance, so Jim Taylor and I went to see Jeremiah Denton. The reception we got was in stark contrast to that Mike Leonard and I had received on our first trip to Washington in support of Flippo's original legislation back in 1982. Then, Denton's staff wouldn't even let us in the office, but took us out in the Senate Office Building hallway and vented their anger at me for criticizing Denton in public. Now, we were being served coffee by the senator's wife and led into the august presence for the first time since the Sipsey fight began.

Denton and his staff were quite cordial. They told us about the confrontation with Heflin in the elevator and Denton said, "He obviously feels very strongly about this. He wants to take the lead, and we'll give him a chance to do it, but if he doesn't, then I'll sponsor the legislation."

Next stop was the office of Senator Heflin, and when we met up with John McMillan and Joe Brown for the appointment, McMillan was gray-faced and somber. "My meeting with him didn't go well. I don't know what we can expect out of Heflin."

Actually, the senator was all smiles and said almost nothing while we outlined the proposal to him. Jim Taylor perceived that Heflin had "oh, hell, written all over his face," since the lack of a compromise was no longer an excuse for inaction. Heflin said only, "I think we're getting there," and that he wanted to be sure of Bill Nichols's agreement to include Dugger Mountain in any legislation. Then, by prearrangement, Jim Taylor and I left John McMillan and Joe Brown there to deal with him.

Jim and I paid a courtesy call on Congressman Tom Bevill, since the gentlemen's agreement provided for restoration to the wilderness proposal of the 1,000 acres in Winston County that Flippo had deleted from his first bill. Bevill was all cordiality, happy to know an agreement had been reached. When we met again with John McMillan, he was feeling better about Heflin.

Next, McMillan, Brown, and I visited Bill Nichols. We were met by a somewhat subdued Winston Lett, and Nichols was not particularly pleased to see us. Turning to me, he said, "Now let me ask you this. Isn't it true that when you asked me for the Cheaha Wilderness you told me that you wouldn't ask me for any more wilderness in my district?"

"Yes, sir, that's true."

"Then what are you doing here?"

"I, er, uh, well, it was the timber industry's idea," I said, pointing to John McMillan. John hunched down a little in his chair and mumbled, "We just didn't want the Bankhead loggers to bear all the brunt of the Wilderness designation; . . . we want to spread the impact around the state a little."

Nichols looked at him and smiled faintly. "Well," he said, "we're all still friends here. I'll take a look at it and talk to Flippo and Heflin. I'll let you know."

The final task was for Jim Taylor and me to brief Ronnie Flippo on all our meetings. When he heard that both Heflin and Nichols had been noncommittal, he said to me, "*Mr.* Randolph, why are you here again without something that Heflin will support?" Growing angry, he started thumping on his desk. "What does Dugger Mountain have to do with the Sipsey Wilderness? I don't care about Dugger Mountain and I don't want Bill Nichols dragged into this. Why should I sign off on this if Heflin won't? Why should I stick my neck out again? I'm not going to go first—I'm not going to do this unless Heflin shows me first that he'll do it."

All we could do in the face of this was assure him that Brown and McMillan were working hard on Heflin and Nichols, and say, "Thank you, sir," and go home.

So, as 1985 came to a close, the fate of the Sipsey expansion seemed to lie in the hands of Congressman Bill Nichols. Winston Lett, having been chastised by Nichols for saying that the Dugger Mountain proposal was fine by the congressman, told the press that Nichols is "keeping his options open. He hasn't said 'No' to the proposal yet, and will study it very carefully."[29]

CHAPTER TWENTY

Might As Well Move

JEREMIAH DENTON WAS NEVER AGAIN a factor in the Sipsey expansion effort. He never introduced legislation of his own. In 1986 he was defeated for reelection by then-Democratic congressman Richard Shelby of Tuscaloosa, in a race in which the Sipsey was not an issue. However, it would be a mistake to overlook or discount the role Denton played. Although his early involvement—or lack of it—merited criticism, his decision to sponsor the negotiating sessions of 1985 forced the timber industry to finally take a specific position, leading them to actually enter into an agreement with us and effectively compelling Howell Heflin to at last give the matter his attention. Recognizing that wilderness preservation is first and last a political process, those who value conservation should recognize that Denton, Richard Lee, and the rest of the staff made a critical contribution to the success of the Sipsey enlargement.

My own situation changed during 1986, as well. For a variety of reasons, I decided not to again seek funding for the Audubon Society's Natural Area Preservation Project. Instead, the Birmingham Audubon Society elected me its president, and it was in this capacity that I concluded the Sipsey effort, as a volunteer. That year I joined a large Birmingham business law firm, and I discovered, interestingly enough, that the little public notoriety I had achieved as leader of the expansion fight had been a positive factor in my recruitment.

Mike Leonard also experienced a career change in 1986. He was recruited by a major regional law firm in Winston-Salem and returned to his native North Carolina. However, Mike still had projects in Alabama that interested him, notably his effort to connect the Pinhoti Trail to the Appalachian Trail, and he was often in the Washington, D.C., area for this and other reasons. Thus, he continued to assist our Sipsey efforts from time to time.

Heflin, meanwhile, seemed less than pleased to find the Sipsey expansion issue now squarely in his lap. On January 21, 1986, he called a group of reporters to his Washington office and told them, "I want to get the thing over with. It seems to be that 90 percent of the agreement has been reached. Might as well move."[1]

The *Mobile Press-Register* reported:

> U.S. Senator Howell T. Heflin suggested Tuesday that the overwhelming majority of Alabama residents don't care whether the Sipsey Wilderness Area is expanded, but he predicted that Congress will pass a compromise plan by this summer anyway.
>
> Probably no more than 500 people really favor enlarging the protected area in the Bankhead National Forest, Heflin half-jokingly told reporters. Another 500 may oppose it, he said. Most of the expansion's proponents, Heflin contended, are newspaper editors or reporters. "The rest of them don't care," said Heflin, D-Tuscumbia.
>
> As for his own feelings, Heflin said, "I like the idea of doing it [creating wilderness areas], but I don't think you should go crazy over it." He blamed environmental groups for the four-year dispute over the proposed Sipsey expansion, saying they tried to undercut the U.S. Forest Service by pushing their own plan before the agency could issue its recommendations.[2]

One of the reporters who attended Heflin's conference was Randy Quarles, Washington correspondent for the *Mobile Press-Register* and the *Huntsville Times*. He was struck by Heflin's hostility, and wrote about it in an analysis titled "'Wilderness-Schmilderness' Attitude about Sipsey Troubling:"

> [C]omments Heflin made to reporters last week may trouble some people concerned about preservation. Not so much the words themselves, but the wilderness-schmilderness attitude they reflect. . . . Heflin made no secret of the fact that he feels the Sipsey fray has been silly. Other than a few hundred folks on either side, he said, the people of Alabama don't give a hang about it one way or the other. . . .
>
> Heflin's observation that few people go to the Sipsey implies that his enthusiasm for the expansion might increase with the number of voters—that is, visitors. That seems an odd criterion for judging the desirability of a wilderness area. By definition, a wilderness is a place where there aren't many people. . . .
>
> The fact that a wilderness is indeed a wilderness inevitably prevents or discourages many people from traveling through it, because such travel is difficult. Although this unfortunately means that many people can't enjoy it, the alternative is to put in roads and parking lots and restroom facilities and an observation tower, complete with elevator, so they can see where the wilderness used to be. . . .
>
> One reporter who hiked along the Sipsey as a Boy Scout years ago . . . remembers

the things that weren't there: no buzzing chain saws or trail bikes; no motor homes with TV antennas and smelly exhausts; no snack bars; no game rooms; no dead possums sprawled on pavement. Those things have their place, just not in wilderness. The retired Boy Scout may never return to the Sipsey . . . but like the white-tailed deer or . . . a painting by a great master . . . it's nice to know the beauty remains.[3]

Before Heflin would "move," Bill Nichols had to make his decision about Dugger Mountain. He and Winston Lett toured the area with Mike Leonard, Carolynn Carr (a Sierra Club activist from Auburn), and Cleburne County commission president and probate judge Mac Smith. Both the *Talladega Daily Home* and the *Anniston Star* published editorials urging Nichols to support the new wilderness area, but in the end, Nichols said no. Ostensibly, the reason given was that Judge Smith opposed the designation, since it might mean the loss of some $2,500 annually in payments the Forest Service made to Cleburne County in lieu of property taxes. (Instead of paying property taxes, national forests made payments to counties based on the number of acres of federal land in those counties. This amount was supplemented by a percentage of the revenue generated by logging and other activities each year. At that time, the Forest Service estimated that Cleburne County might receive $2,000 to $2,500 less each year if Dugger Mountain were designated a wilderness.) Nichols also said he was "unsure whether another large wilderness area is needed just 25 miles north of the Cheaha Wilderness Area."[4]

Mike Leonard feels that Nichols simply didn't want to do it: "I thought I was done with this wilderness issue," he told Mike. "I don't like being brought back into it." Ronnie Flippo's ambivalence toward Dugger Mountain must also have been a factor, but my own impression is that Nichols was still a little miffed at having the Cheaha Wilderness forced upon him, as it were, at the expense of his Talladega Scenic Drive project. In any event, nothing was to be gained by criticizing Bill Nichols, so we decided to wait and see what Heflin would do.

Meanwhile, the U.S. Forest Service released its final Alabama forest plan, to take effect April 20, 1986. Its wilderness, wild and scenic river, and semi-primitive recommendations roughly mirrored the last proposal Richard Lee and Joe Brown had devised during Senator Denton's negotiating sessions the previous year.

For the first time since the inception of the Sipsey campaign, Heflin did not make either Congressman Flippo or the Wilderness Coalition learn of his intentions by reading about them in the newspaper. In early May he distributed his draft Sipsey Wild and Scenic River and Sipsey Wilderness Addition Act of 1986. These were the highlights:

1. Designation of the West Fork Sipsey and all but one of the upper tributaries as units of the National Wild and Scenic River System. The excluded tributary was Montgomery Creek, where the Forest Service was "directed to construct a dam to establish a substantial lake for all types of recreational use . . . in the Montgomery Creek watershed" or in some other location in the Bankhead "if the Montgomery Creek watershed area is not suitable." The Forest Service was also "authorized and directed to construct a paved road" to the new lake.
2. A puzzling mandate upon the Forest Service to install water quality monitoring devices at the southern end of the wild and scenic river, and "to take all actions . . . necessary to eliminate any conditions causing injurious water quality." Heflin had apparently included this to counter criticism that he was indifferent to protecting the water quality of Smith Lake and its sources of municipal water supply. However, the wording of this section seemed predicated on the notion that preserving the river would be the cause of water degradation rather than its prevention.
3. Enlargement of the Sipsey Wilderness Area by "9,990 acres," an inexplicable deviation from the acreage in the gentlemen's agreement that I am unable to account for to this day. However, the proposed expansion still included the watersheds of Hagood and Braziel Creeks, and part of Borden. Dugger Mountain was not included.
4. Creation of a "Bankhead National Recreation Area" comprised of the "Wilderness Area and the Lake area with sufficient surrounding area to make an adequate recreational area." The Forest Service was to "administer the Recreation Area in accordance with . . . regulations applicable to the National Forest System," meaning that the Forest Service would have wide latitude in deciding activities permitted there, although it was directed to "reduce the use of 'clearcutting' as a silvicultural tool where possible."
5. Direction to the Forest Service, "[n]otwithstanding the provisions of . . . the Wilderness Act . . . [to] take such actions as deemed necessary to eradicate the Southern Pine Beetle within the Sipsey Wilderness, Sipsey Wilderness Additions, and adjacent Bankhead National Forest . . . and thereafter [to] take pre-

ventative maintenance measures to control the spread of the Southern Pine Beetle."

6. Instruction to the Forest Service to "allow public access, including the use of motorized vehicles, to the existing roads within the Sipsey Wilderness Addition."[5]

The road openings and the direction to eradicate and control the pine beetle meant that Heflin was proposing that the Sipsey Wilderness be managed in a manner that contravened the Wilderness Act and that it be coupled with an unnecessary "recreation area" and a pork-barrel lake project, to boot. Surely the senator knew he could never get this passed by the Congress.

"We are concerned that many aspects of the bill will attract national opposition and will be extremely difficult to get approved, particularly in the House," said [John] Randolph. "We ask that if Sen. Heflin comes to the same conclusion, he sever these extra matters so the basic wilderness and wild and scenic river proposals can be resolved. . . . " John McMillan, executive vice president of the Alabama Forestry Association, said . . . he agreed with Randolph that the new provisions would likely run into congressional opposition.[6]

Heflin told the press that his draft was not final and that he expected "modifications, additions and deletions" to be made. What most concerned us, however, was that 1986 was the second year in the 99th Congress, which was targeting adjournment in October. If Heflin was serious about passing a bill that year, he would need time to fight through all the extraneous matters he had raised. There needed to be immediate action if he truly wanted to "get the thing over with."[7]

May and June came and went with no movement out of Heflin, and national conservation groups began to weigh in on some of the details of the proposed legislation. Motor vehicles on the dirt roads within the expansion area drew instant objection. "'One of the main things one wants to accomplish with wilderness designations is to keep motorized vehicles out of them,' said David Alberswerth, legislative representative for the National Wildlife Federation in Washington. 'I'm fairly confident in saying that if the intent is to have continuous motorized use of roadways within the proposed wilderness boundaries, I don't think that that's going to fly—in the House of Representatives, anyway.'"[8] Michael Scott, the Wilderness Society's deputy director for conservation, said, "Allowing motor vehicles in the Sipsey Wilderness would be 'antithetical to the whole concept of wilderness, [where] you see nature in its primeval state, not from a jeep'."[9]

Of course, Flippo's original legislation had provided for the Northwest Road to be left open, but that was for a wilderness area expanded by nearly 28,000 acres to a total of about 40,000 acres. Heflin's proposal of just under 10,000 acres would include only some three miles of the Northwest Road. Joe Brown was telling everyone that the Forest Service would prefer that this stretch be closed and used instead as a horseback trail. Heflin's proposal implied that not only the Northwest Road, but *all* "existing roads" be opened to motorized use, conceivably including old logging roads and historic tracks that ran along several of the divides.

The dam project also raised eyebrows. "Heflin's call for construction of a dam to create a lake . . . near a stream that would be part of a 'wild and scenic river' system, also could be controversial, said Eric Olson of the American Rivers Conservation Council. The federal Wild and Scenic Rivers Act currently forbids federal projects that would adversely affect the flow of a wild and scenic river, Olson said. 'I think that will be a source of major debate.'"[10]

By July, with the clock ticking on the congressional session and still no Senate bill, Heflin was telling the press that these objections could be "worked out," and that the most important issue to him was pine beetle control. "'What we want to do is to get the Forest Service to take on the responsibility' to solve the problem. [Heflin] said he wants to see language in a bill that would allow the Forest Service to 'go in and be able to treat the condition pertaining to the pine beetle and hopefully eliminate it. . . . [Y]ou've got a problem with the national environmental groups on the issue of what the Forest Service can do pertaining to the pine beetle. . . . I'd say that's the biggest issue that remains unresolved.'"[11]

To understand the ensuing controversy over the pine beetle issue, one must understand how pine beetles are controlled. Heflin's implication that the problem could be "treated" conjures up visions of low-flying aircraft laying a fog of pesticide over the forest, or perhaps, like an army of Orkin men, forest rangers spraying the base of infested trees. This is not how it's done. The pine beetle cannot fly; it hops, crawls, or glides on the wind from tree to tree. It is controlled in two ways, both of which involve logging. An active infestation is "treated" by clear-cutting a large enough perimeter around an infested area—healthy trees as well as diseased—so that the little booger cannot get a fresh supply of food; then it dies. If cut early enough, infested trees can be salvaged for commercial purposes and sold along with the healthy ones. Prevention, or control, involves periodic cutting of young pines (called "thinning") so that the remaining pines have the best chance to grow into healthy maturity, as pine beetles tend to attack the weakest trees.

Heflin, in short, was proposing to open portions of the Sipsey Wilderness Area for commercial logging on a continuing basis.

Finally, on August 15, 1986, Heflin introduced his legislation, S. 2782, with a few changes from his original draft. The expansion acreage was up to 10,070 acres. Gone was the "Bankhead National Recreation Area," and the site of the lake, now specified to be 1,000 acres, was moved out of the West Fork Sipsey watershed to "either Brushy Creek or Flint Creek," with a standard national forest recreation site adjoining (campground, picnic area, boat launch, etc.). Only the Northwest Road was to be left open for motor vehicles. Otherwise, the bill was the same, including the pine beetle language.

Announcing the bill, Heflin said:

Some individuals have labeled me as the obstructionist in the Sipsey Wilderness expansion process, but I want to make it abundantly clear that I support the wilderness concept. Our wilderness areas are, in fact, showplaces and gymnasiums of nature—places for scenic, leisure, refuge, recreation, and spiritual and physical refreshment....

To state that the Sipsey Wilderness expansion proposal is controversial is a gross understatement.... [It] has been the subject of emotional debate almost from the time the issue surfaced in legislative form. Unfortunately, the proposal became the forum where the views of the tree huggers versus the tree killers were acted out....

Somewhere along the line of this controversy, the focus was centered on the two extremely divergent views and the overall public interest was seemingly shunned. The idea of considering different approaches to sound use of these precious natural resources and sound environmental protection was forgotten....

I had hoped to have a complete agreement on this proposed legislation before it was introduced. However, there are still many differences. Since the time is running out on this session, I am introducing a proposal that can be amended."[12]

In a press release a few days later, Heflin asserted, "We have plenty of time to work this out before the end of the 99th Congress."[13]

CHAPTER TWENTY-ONE

It Cannot Be

Now it was Ronnie Flippo's turn to surprise everyone. On September 12, 1986, he introduced his own version of a Sipsey expansion, this one for just under 16,000 acres. Included were the upper and eastern portions of the Thompson Creek RARE II area, including both Thompson Creek and Mattox Branch (but omitting Tedford Creek); the entire watersheds of Hagood and Braziel Creeks; and part of Borden. All the wild and scenic river designations were included. "[Flippo] noted that he pushed bills through the House in 1982 and 1983 to add 28,000 acres to the area. 'So why would I now not want more [than Heflin's bill]?' Flippo asked.... 'All matters of the Sipsey bill are proper subjects for negotiations,' Flippo said.... 'I think everything is negotiable.'"[1]

Heflin, of course, was miffed: "'It was represented to me by the environmental groups that all parties involved had agreed to the acreage pertaining to the expansion of the wilderness,' said Heflin.... 'From my conversations with Congressman Flippo and his public statements, I now find these representations made to me were in error because a major player, Congressman Flippo, had not agreed to the so-called compromise,' said Heflin."[2]

The senator was mistaken. We had made no "representations" to him—in fact, he rarely sought our input—and we certainly never told him that we spoke for Ronnie Flippo. Flippo himself had been both publicly and privately refusing to take a position for nearly two years until he saw what the senators would do. Perhaps if Howell Heflin had ever actually involved Ronnie Flippo in a meaningful way, this situation might not have arisen.

As for the Wilderness Coalition's participation in any compromise, the only agreement we had reached was the gentlemen's agreement of 1985, which provided for an 8,400-acre Dugger Mountain Wilderness in addition to the Sipsey enlargement. We did urge Heflin to "proceed with such a final version of the legislation as you feel comfortable in supporting," offering our "complete cooperation in working within the legislative process to help you bring this issue to a rapid and amicable conclusion."[3]

An article in the *Birmingham News* reported on reactions from Flippo and Denton:

> Flippo said he would like to see more acreage included than was set out in Heflin's measure. Flippo said . . . he spoke with Heflin, but the senior senator apparently would not compromise. "He wants to continue to have hearings on his bill and I think that's a good thing," said Flippo. "We are running out of time in both the House and the Senate so I'm hopeful that he will go forward with his hearings and I certainly intend to go forward with hearings on this side." . . .
>
> A spokesman for [Jeremiah] Denton said the senator would support Heflin's bill as a vehicle to move the issue through the Senate, but prefers the increased acreage contained in Flippo's measure.[4]

Flippo's bill was heard before the House Subcommittee on Public Lands on September 18. Besides the acreage difference, the proposal also varied from Heflin's in that it did not contain the pine beetle language. However, the provisions for monitoring water quality and for the new lake and recreation area were included. Flippo said these provisions had "only been included in the first place to make the bill compatible with [Heflin's]," although Flippo also said he had been urged by constituents in his district, including the mayor of Moulton, to support the new lake.[5] He had included it, at least in part, to satisfy them.

F. Dale Robertson, deputy chief of the U.S. Forest Service, essentially opposed the entire bill, that is, acreage greater than that recommended by the Alabama forest plan, the unnecessary burden of water quality monitoring, and, most especially, the unwanted lake, which he estimated would cost between $18 million and $22 million. The only thing the Forest Service did support was the wild and scenic river designations.

The *Florence Times-Daily* reported:

> Representative John Seiberling, D-Ohio, chairman of the House public lands subcommittee, struck the proposed lake and protection of the Sipsey River's water quality on a voice vote. . . . Seiberling said the water quality provision restates existing law and is "confusing."
>
> The subcommittee's opposition to the recreational lake was expressed last week by both Seiberling and Rep. Buddy Darden, D-Ga., who said linking the lake project to

the wilderness expansion could be "fatally defective." When Seiberling said he would propose a substitute for Flippo's bill, Darden demanded to know "is the $22 million dam still there?"[6]

Cleared of everything but the wilderness expansion of 16,000 acres and the wild and scenic river designations, the bill was favorably reported out by the subcommittee to the full House Committee on Interior and Insular Affairs, which, a few days later, approved it and sent it to the House floor, where it passed on a voice vote.

Now it was Heflin's turn. Sen. Jesse Helms of North Carolina, chairman of the Senate Agriculture Committee, extended to Heflin the courtesy of letting him chair the hearings on his own bill, which were held on September 25, 1986. In his opening statement, Heflin made it clear that he was standing his ground on the pine beetle issue:

Let me just say that I am not going to be responsible for a congressionally mandated, 10,000-acre Southern pine beetle refuge, so the infamous critter can freely go about destroying the natural treasures that we are seeking to preserve in this legislation. . . . Alabama has already lost thousands of acres of prime forest land to the pine beetle, and if . . . steps . . . to control the spread of this pest . . . are not taken, the pristine wilderness areas, as they currently exit, could be lost forever. . . . My bottom line is that I am not going to support a wilderness bill unless it has provisions in it dealing with the southern pine beetle . . . authorizing clearly the Forest Service to take actions to minimize and to take reasonable steps to try to eradicate it in the existing Sipsey Wilderness and any expansion [of] it.[7]

As logical and straightforward as these statements might seem, they were, in fact, misleading.

The 10,000 acres he was referring to contained the watersheds of Hagood and Braziel Creeks, and part of Borden, a land characterized by steep slopes, rock-walled canyons, and a predominance of hardwood trees. Only along the divides between the canyons, most particularly on the periphery of the RARE II areas, where road building had been feasible, were there scattered stands of pine plantations. Pine beetle infestation there would not "destroy the natural treasures" of the Sipsey Wilderness, which lay in its old-growth hardwoods and canyon ecosystem, but only the pine plantations. Over time, the natural forces that wilderness designation is intended to encourage would regenerate those areas to the balance that nature intended for this environment—mixed pine and hardwoods on some sites, but predominantly upland hardwoods. From this perspective, a pine beetle infestation is not necessarily a bad thing.

In fact, the pine beetle offers a validation for national forest wilderness preservation. Recall that the eastern wilderness movement was born of outrage over widespread conversion of natural old-growth forests into plantations of pine. The devastation the pine beetle can wreak in areas that have been manipulated into a pine monoculture underscores the foolishness of allowing such a practice in the first place, especially in publicly owned national forests, and most especially in the small ones in the East, which, in law if not in practice, are supposed to be managed for uses other than just commercial timber production.

But Heflin seemed indifferent to all this. The pine beetle offered a means to permit logging within the Sipsey Wilderness Area. This is what SWUFFL wanted. For years, loggers had complained bitterly whenever they were prevented by the Wilderness Act from removing timber damaged by a tornado or pine beetle infestation in the Sipsey.[8] Heflin appeared to be responding directly to these complaints.

Deputy Forest Service Chief Dale Robertson objected: "We [the Agriculture Department and Forest Service] do not support [the sections] of the bill which would require the Secretary [of Agriculture] to endeavor to eradicate the Southern Pine Beetle from the Sipsey Wilderness and adjacent Bankhead National Forest. Adequate authority exists for the Secretary to control the spread of Southern Pine Beetle outside the Sipsey Wilderness. Within the wilderness, periodic thinning of young pine forests would be incompatible with wilderness designation." In fact, there was very little of Heflin's bill that Peterson would support:

> We strongly oppose . . . the [required] construction of a lake of 800 to 1,000 acres within the Bankhead National Forest for the sole purpose of recreation. The estimated $20 million cost of this lake and related facilities such as access roads, campgrounds, and boat launching ramps is not justified in view of other priorities and the Federal budget deficit. Furthermore, we have not done any feasibility or environmental studies to evaluate such a project. . . .
>
> We also urge deletion of the [provisions requiring] the Secretary of Agriculture to . . . monitor the quality of water flowing into Smith Lake and to take actions needed to control any conditions causing injurious water quality. Some of the waters in the Sipsey Fork flow from sources on private land which are beyond the control of

the Secretary of Agriculture. . . . The Secretary already has authority to correct conditions which are causing resource damage within wilderness areas. . . .

We strongly oppose [the provision allowing] the use of motorized vehicles within the proposed Sipsey Wilderness Addition. The use of motorized vehicles is very incompatible with wilderness and specifically prohibited by the Wilderness Act.

Robertson also quibbled with the acreage in Heflin's proposed expansion, preferring that the bill conform to the Alabama forest plan recommendation. In other words, as had been the case with Flippo's legislation, the Forest Service was only supportive of the wild and scenic river designation.

The next witness was Ronald J. Tipton, who had replaced Randy Snodgrass as the Wilderness Society's southeast regional representative. After making a statement supportive of the acreage in Flippo's bill over that in Heflin's, Ron Tipton engaged Heflin on the pine beetle issue. Noting that the Wilderness Act already contains language permitting the Secretary of Agriculture to take "such measures . . . as may be necessary in the control of fire, insects and diseases," he described two wilderness areas in Texas and Louisiana that the Forest Service logged under this authority in an effort to control the pine beetle.[9] Unlike the Sipsey, with its predominance of hardwoods, these two preserves featured large natural stands of coastal plain pine forest. The logging that was permitted there was extensive enough to have produced litigation challenging the practice, with the result that the Forest Service was in the process of developing a completely new policy on pine beetle control in the Southeast, the draft of which was then in public circulation.

Mr. Tipton: . . . The concern that we have is based on experience in states like Texas and Louisiana, where large areas of existing wilderness have been cut, and it is not clear that is going to be effective in controlling the spread of this beetle. We have a real problem with cutting down the small designated wildernesses we have in these Southern States for the purpose of trying to control a natural pest that is very difficult to control. We are afraid that . . . [Heflin's] language would have the effect of superseding pending litigation and the Forest Service's own [pending new policy] on pine beetle control in the Southeast. . . .

Senator Heflin: Well, Mr. Tipton, I really see the Wilderness Society and the national [environmental] groups . . . [as] being crucial to whether or not we pass a Sipsey bill before the end of this Congress. If you continue your opposition to efforts to try to put language in the bill pertaining to the southern pine beetle, then I see it [causing] a problem and really probably means that it cannot pass.

Mr. Tipton: . . . I think there is another party to this that is probably more important than the Wilderness Society or the Sierra Club, and that is Congressman John Seiberling, chairman of the Public Lands Subcommittee in the House. Congressman Seiberling has had hearings on . . . controlling the pine beetle in wilderness. He is very concerned about what has happened in Texas and . . . Louisiana. . . . I think he would not want to give the Forest Service carte blanche authority to take action that is deemed necessary to control the beetle, or to waive existing laws. . . .

Senator Heflin: . . . Now, you can get somebody, a Senator or Congressman, to espouse your viewpoint and oppose it, and you can hide behind their coattail . . . get somebody else to do your dirty work . . . but I am just saying that in the shortage of time that we have that remains, if you want this passed at this time, it is going to have to be done with agreement. . . . [I]f we are going to pass this wilderness and pass it now, it has got to have complete, 100 percent agreement [of] all parties concerned. Otherwise, it cannot be.

The exchanges with these witnesses comprised the meat of the hearing. Others appeared, of course, including Jim Taylor for the Sierra Club, and me, representing the Birmingham Audubon Society. Paul Kittle of the University of North Alabama in Florence appeared on behalf of the Shoals Area Audubon Society. SWUFFL's Bill Bustin testified in favor of Heflin's bill over Flippo's. And a voice from the past, Charles Prigmore, the primary lobbyist in the original Sipsey Wilderness fight, appeared essentially on his own motion to encourage "compromise and conciliation," saying, "We in Alabama . . . are perhaps best equipped to decide what should be done in the expansion of the wilderness in Alabama. I do not think we should pay more than perfunctory attention to the positions of national organizations here."

Prigmore, like Heflin, seemed unwilling to recognize the opposition the pine beetle language would provoke in the House of Representatives. As for the Alabama Wilderness Coalition, we were keeping our yaps shut, but there was no way that our groups were going to agree to let the Sipsey Wilderness be managed as anything less than a unit of the National Wilderness Preservation System; nor would we consent to expansion legislation that served as a precedent for weakening that system.

After Ron Tipton agreed to talk with Congressman Seiberling to determine if there was any alternative pine beetle language he could propose that would not contravene the Wilderness Act, Heflin adjourned the hearing.

Ron recalls the experience as "one of most unusual hearings I've ever been to, first of all because it was so informal."[10] Instead of a large hearing room with members of Congress seated at a dais at the head of the room, looking down on witness tables in front, the Senate Agriculture Committee met in something like an oversized conference room dominated by a rectangular table. Heflin sat at the head, and the witnesses sat around the table with him. Second, said Ron, he was struck by "Heflin's being alternately pseudo-charming and being a bully. It really felt more like a meeting of the parties in interest than it did a [congressional] hearing." Which, of course, is exactly what it was.

A few weeks later, Heflin was back in the press:

U.S. Senator Howell Heflin, D-Ala., lashed out at a national environmental group . . . blaming it for holding up passage of legislation to expand the Sipsey Wilderness Area. . . .

"The National Wilderness Society has commitments from several members of the House and Senate to conduct delaying tactics which will kill the bill if it contains language undesirable to them," said Heflin. The senator's press secretary, Jerry Ray, could not say who Heflin was referring to.

"That's simply inaccurate; it's not true," responded Ron Tipton, Southeastern regional representative of the Wilderness Society in Atlanta. . . .

"It comes down to the Southern pine beetle," said Heflin. "The national wilderness groups don't want the Sipsey bill to have any language which modifies the general wilderness law," he said. . . . "We have worked out satisfactory language with the U.S. Forestry Service [*sic*] on this issue," said Heflin. "If we can work out satisfactory language . . . with the National Wilderness Society, I believe we can still pass a bill. . . . "

F. Dale Robertson, associate chief of the Forest Service, said . . . he knew of no specific agreement with Heflin over language. He testified at a recent Senate hearing . . . that the Forest Service opposed Heflin's language regarding the pine beetle because it calls on the agency to "eradicate" the bugs, an impossible task.[11]

Ronnie Flippo had this to say about the stalemate: "'The key is for the senators to act on a bill.' . . . Flippo said the method of resolving legislative disagreements—'proven over 200 years'—is for the House and the Senate to each pass a bill and to settle the differences in conference. 'The house has passed a bill; the Senate has not,' said Flippo."[12]

Congress adjourned in mid-October. Heflin made no effort to pass his legislation, the only "delaying tactics" being his own. He left the bill lying in committee, never asking that it be approved and sent to the Senate floor for a vote.

Heflin blames the environmentalists for the stalemate, but Alabama and national environmentalists say the senator is the reason the 99th Congress adjourned . . . without taking final action. . . .

"I don't think he intended to move this bill," said Ron Tipton, Southeastern regional director of the Wilderness Society in Atlanta. "The issue is not the Southern pine beetle. The issue is whether he wants an Alabama wilderness bill."

Heflin was working with the Wilderness Society on trying to come up with language that both sides could accept. The Tuscumbia Democrat said the society "never got back with me" on a compromise. Tipton, however, said Heflin made only slight changes in his proposal, leading the society to conclude that further discussion was "a fruitless exercise."

[Rep. Ronnie Flippo] said that Heflin's concern with pine beetle language is a "non-issue," and expressed hope that the matter could be ironed out next year. . . .

John Randolph of the Birmingham Audubon Society said . . . "We've spent more than two years intensively trying to do everything Senator Heflin demanded of us. . . . We handed the issue to him on a silver platter by reaching an agreement with our opponents. Yet he always comes up with some other reason on why he can't get it resolved."[13]

CHAPTER TWENTY-TWO

Peace Is at Hand

THERE HAD BEEN LITTLE CAUSE for levity in the long Sipsey campaign, so a piece written in January 1987 by the *Birmingham Post-Herald*'s Washington correspondent, John Brinkley, offered a welcome diversion:

Sipsey Deadlock Loses Dependability

[T]here are here in Washington certain events that happen on a routine, regular basis—especially at the start of a new year. There is, for example, the advent of a new federal budget cycle . . . the president's State of the Union address . . . a new political campaign season . . . [and in] odd years, January brings a new Congress.

And, of course, every new year brings with it that most predictable of occasions—the start of another battle in the War of the Sipsey Wilderness.

The regularity of these events . . . has a soothing effect upon those of us who depend on [them] . . . to retain a semblance of order in our lives. But danger lurks. Toward the end of the 1986 round of Sipsey combat, it appeared that the war might actually come to an end. Yes, it's true. The matter was nearly settled once and for all. . . . Fortunately, though, rational minds prevailed. . . . [The] participants . . . did what had to be done. They saw to it that no agreement could be reached. . . . The 99th Congress adjourned with [Howell Heflin and Ronnie Flippo], thankfully, unable to reach a compromise and visibly displeased with each other.

But those of us with some interest in a perpetuation of this conflict can see that the differences between the two proposals were not drastic [and] . . . with a little more time could have been resolved. . . . This is cause for alarm.

Here's what you can do. First, write to Congressman Flippo. Tell him that the Sipsey should not be enlarged by 15,000 acres—it should be 15 million. At least. Then tell him that the pine beetle should not be eradicated; it should be declared an endangered species and given all possible protection. . . .

Next . . . write to Sen. Heflin. Tell him if he really wants to protect some woods, OK, but an area about the size of a tennis court. . . . And tell him to put language in his bill authorizing—no, ordering—the U.S. Air Force to wipe out the pine beetles by dropping a neutron bomb on the forest. It would kill the insects but leave the trees standing. . . .

The urgency of this issue cannot be emphasized strongly enough. There is a clear and present danger that the Sipsey Wilderness conflict will be resolved in 1987, thus depriving us all of what has come to be a rock of stability in our lives.[1]

Would that Brinkley's prediction had been true, but as he noted, Heflin was "visibly displeased." After the 100th Congress convened in 1987, Mike Leonard happened to be in Washington and discussed the situation with Heflin's staff. The senator had directed his attention to other things, and the staff was unsure when or if Heflin would return to the Sipsey. But they knew that Heflin would not sponsor any new legislation "with Marshell Frost lying there on his death bed."

SWUFFL's founder and president, Marshell Frost, had cancer and would not survive the year. I hardly knew the man, but I had been around him enough to know that he was sincere in his beliefs. Like Charles Borden, he had been raised in the Bankhead Forest area. He had helped construct the Bunyan Hill Road, one of the dirt tracks that would be closed by the wilderness expansion, and had earned his living from the national forest all his life. He was a plainspoken opponent of wilderness designation. Didn't believe in it, period. He proved to be mistaken in his fears of harm to the local timber industry, but fear can be a powerful motivator, and, as I say, he was sincere.

While Heflin certainly may have been waiting out of respect for Frost, more likely he was simply delaying action until the end of the congressional session, as he had done in 1986. Waiting until the last days of the 100th Congress would increase pressure on the parties to accept a Sipsey enlargement on Heflin's terms.

Toward the end of 1986 I had chaired the annual meeting of the Alabama Audubon Council at Dauphin Island. Ronnie Flippo and Ron Tipton were both invited speakers. Tipton says the meeting particularly stands out in his mind because of an announcement I made at 1:00 P.M.: "You said, 'We're now going to break for the Alabama–Notre Dame football game and relive the War of Northern Aggression.' I was still [learning] the culture of the South. I [had] never been to an environmental meeting—and I've been to a lot of them—

where they stopped the meeting in order to watch a football game. That was a first."

At Dauphin Island, we had a very cordial meeting with Ronnie Flippo, in which he reaffirmed his commitment to Sipsey legislation but still wanted to wait for Heflin. He felt confident he could make the legislative process work this time, and we agreed to leave it entirely in his hands.

For our part, we told Flippo we would shift our emphasis away from trying to satisfy Heflin and would instead attempt to force some new concessions out of the Forest Service by challenging the Alabama forest plan.

In 1986, with guidance on legalities and language from the Wilderness Society and the Sierra Club, we had quietly filed with the chief of the Forest Service in Washington a formal administrative appeal of the Alabama forest plan. Appellants were the Birmingham Audubon Society, the Alabama Chapter of the Sierra Club, the Alabama Conservancy, and the Wilderness Society. While the grounds for the appeal included many points that the national groups were pursuing in other administrative appeals in hopes of reforming the Forest Service's overwhelming preference toward commercial timber production, our appeal specifically challenged the adequacy of the wilderness recommendations in the Bankhead National Forest. The chief's office issued a routine moratorium on timber production in the 30,000 acres of the Bankhead RARE II areas until September 1987, or the decision on the appeal, whichever came first.

I had prepared the appeal document myself as a volunteer for the Audubon Society, but now my duties with my new law firm were requiring my full attention. We needed help if we were to seriously press this appeal and, if necessary, ultimately file litigation.

Birmingham native Rick Middleton had been employed for some time as an attorney for the Sierra Club Legal Defense Fund, which, as the name implies, is the club's legal arm. Recently, he had decided to take advantage of a funding opportunity to create a public interest law firm concentrating on the South's environmental problems. Thus was born the Southern Environmental Law Center (SELC), headquartered in Charlottesville, Virginia, and the Alabama national forest plan appeal became one of the firm's early projects. Rick assigned attorney David W. Carr Jr. to the task. The legal work was to be done for free; the appellants agreed to pay expenses. We announced our intentions to the press: "Alabama environmentalists have enlisted a public interest law firm in an effort to stop any development in potential Sipsey Wilderness expansion areas. . . . David Carr, a staff attorney with the [SELC] said the forest management plan would unacceptably 'release 65 percent' of [the Bankhead RARE II] acreage to development."[2]

"These priceless, natural lands are being sacrificed to development despite the fact that they comprise less than one-fiftieth of the commercial forestland in Alabama," [David] Carr said.

John Randolph said . . . "If a responsible member of the Senate delegation from Alabama wants to resolve the dispute over [the Sipsey expansion], there's no doubt in my mind that it can be done. . . . But in the meantime, we're going to keep all of the land up there out of timber production for as long as is humanly possible. . . . Unless Congress acts soon to resolve the Sipsey Wilderness issue, it appears that this will be just one more Alabama problem to be decided by a federal judge." [3]

Also now stepping in to help was Alabama attorney general Don Siegelman, an early and active supporter of both the Cheaha and Sipsey expansion efforts, and an old college chum of mine. Don assigned the head of his Environmental Division, Craig Kneisel, and a young assistant attorney general, Ray Vaughan, to render whatever help we and SELC might need in the coming fight. We had a good team now, and had done all that we could do, so I turned my attention to practicing law.

The forest plan appeal had its desired effect—it motivated the Forest Service to intensify its efforts to resolve the Sipsey impasse. Actually, Alabama forest supervisor Joe Brown needed little additional incentive. He had been working on the issue in good faith, within the confines of the philosophy of his agency, and his own, since arriving in Alabama. When the effort intensified as a result of Jeremiah Denton's negotiating sessions, he made proposals for solutions on a variety of issues, pushing the process that ultimately led to the gentlemen's agreement in 1985. He wanted, as sincerely as anyone involved in the Sipsey Expansion dispute, to see a legislative resolution, and when the opportunity again arose in the 100th Congress, no one worked harder than Joe Brown to produce the measure that was finally approved. I considered him a friend, and so, I believe, should the people of Alabama.

Joe needed time to work his magic, so when the logging moratorium granted for the forest plan appeal approached its expiration date, he asked for a meeting. We gathered in a conference room at my law firm on July 6, 1987.

David Carr came down from Virginia, and Assistant Attorneys General Craig Kneisel and Ray Vaughan appeared for Don Siegelman. A variety of minutiae was discussed—boundaries for the wild and scenic river corridor, the definition of activities to be permitted in semiprimitive areas, that sort of thing—but the most important agreement reached was for an extension of the timber moratorium on the Bankhead lands.

As for such issues as acreage, the recreation lake and, of course, the dreaded pine beetle, we would leave these for Brown to work out with Ronnie Flippo and Howell Heflin. I told Joe that if he could get the congressional delegation to pass reasonable Sipsey legislation, then the Wilderness Coalition groups would be willing to drop the forest plan appeal.

And so, events crawled on. Nineteen eighty-seven gave way to 1988, the second and last session of the 100th Congress, and the time when, if he would act at all, Heflin could be expected to move.

In June 1988 the *Birmingham News* ran an article on Howell Heflin's record on environmental issues, and, of course, the Sipsey expansion was discussed:

> Ron Tipton, who worked on the Sipsey issue as southeastern regional director of the Wilderness Society, said Heflin "doesn't like wilderness very much, and clearly he doesn't listen to the Alabama environmental community. But I've gotten beyond being mad about it," he added. "I just accept it as being part of life."
>
> Tipton . . . said that he is surprised that Heflin's positions on the environment are so conservative. "He's not a strongly conservative, right-wing person in the whole range of issues. . . . He's considered somewhat of a moderate on many issues."
>
> Tipton said that the Wilderness Society has been able to "deal with conservative people in the Congress on wilderness, and we've dealt with them more effectively than we have with Senator Heflin."[4]

Heflin brushed off the criticism, saying that he expected to introduce a Sipsey bill before the end of the session, for Joe Brown had, in fact, been making progress with him. Ronnie Flippo, for his part, seemed to be growing angry again. Mike Leonard stopped by to see him on another of his Washington trips, and found Flippo "mad as a hornet. He was vehement about Heflin's refusal to help him with the Sipsey." When Flippo was a member of the Alabama legislature and Heflin the state's chief justice, Flippo was one of the prime movers of Heflin's pet project—the thing that essentially made his political reputation—the "judicial article," a legislative reform of Alabama's judicial system. Flippo felt that Heflin owed him for his support in those days, but, says Mike, "It was as though Flippo did not want to raise those issues directly with Heflin and was venting on me, instead. It was odd."

Finally, Flippo did raise those issues with Heflin. Flippo says he also told the senator, "I wanted very badly to see [the wilderness expansion] accomplished, and I wanted to convey the idea that this issue is not going away. As long as we're here, we're going to introduce this bill." When the matter was finally stated in such a direct and personal way, Heflin found it difficult to refuse. "I guess he just got tired of fooling with the issue," Flippo says.[5]

In August, Heflin circulated draft legislation for comment (but not to us—we got our copy from Flippo). There was some concession on acreage, but still included were the dam and recreation site, retention of the Northwest Road for motorized vehicles, water quality monitoring and, of course, pine beetle language. Heflin had made some revisions to his previous wording on the pine beetle, but the language still attempted to amend and contravene the Wilderness Act, and to permit logging within the Sipsey Wilderness to control the "infamous critter." The Wilderness Society said they would oppose it. Once again, we were headed for stalemate and controversy, and the short time left in the congressional session was rapidly becoming a factor.[6]

I was disturbed to receive an angry telephone call from Ronnie Flippo's legislative assistant, Frank Toohey. "John, those Wilderness Society people are dealing directly with Heflin's staff, telling them they won't support this and they won't support that. If you guys don't stop interfering in Congressman Flippo's business, he's going to drop the whole thing!" All I could do was assure him that no interference was intended and that I would get to the bottom of it right away.

Just as I hung up with Toohey, Ron Tipton called. "Boy, I've just been reamed out by Frank Toohey and he's threatening to drop the Sipsey bill. I thought I was helping. I sure don't want to be the cause of any problem with Flippo."

Actually, a member of Heflin's staff had initiated contact with Tipton, trying to convince him to accept yet another version of pine beetle language. Since this one still called for logging in the wilderness, Tipton had told him it would be unacceptable, and the Heflin staffer had complained to Frank Toohey. We resolved the tiff with the promise to Frank that anything the Wilderness Society had to offer would be funneled through Congressman Flippo. Ron recalls now that Toohey "did sort of shake me up. You never can tell when somebody representing a member of Congress threatens you, in effect."

Finally, in September 1988, Joe Brown reported that Heflin was relenting and was now willing to let the legislative process work on the pine beetle language and other extraneous provisions. The key to this concession was that the Forest Service had finally adopted its new comprehensive policy toward pine beetle control in the South, which pacified Heflin somewhat. Assuming that an accommodation could be reached on the peripheral issues, only the acreage to be included in the wilderness expansion remained unresolved.

Flippo and Heflin met once more and reached an agreement at last. The expansion would include the southeastern third of the Thompson Creek RARE II area, the complete watersheds of Braziel and Hagood Creeks, and part of Borden—essentially the same land, ironically, that had been included in the Montgomery Plan devised by Joe Brown and Richard Lee of Senator Denton's staff back in 1985. After Congressman Tom Bevill consented to the addition of the land in Winston County that Flippo had removed from his original bills, the total acreage was 13,260, more than doubling the size of the Sipsey Wilderness Area. An additional 5,000-plus acres outside the wilderness would be included in the National Wild and Scenic Rivers System (see map p. 191).

Now that the heart of the controversy had been resolved, Heflin allowed negotiations between his and Flippo's staffs with the Wilderness Society, the Forest Service, and the staff of the House Public Lands Subcommittee to produce an agreement on all the other matters, including the pine beetle, leading John Brinkley, Washington correspondent for the *Birmingham Post-Herald*, to declare, "Peace is at hand in the war over expansion of Alabama's Sipsey Wilderness."[7]

And though it seemed hard to believe, indeed it was.

On October 28, 1988, Pres. Ronald Reagan signed into law P.L. 100-547, the Sipsey Wild and Scenic River and Alabama Additions Act of 1988, and a Sipsey Wilderness expansion of 13,260 acres became reality.[8]

Also enacted was a small enlargement of the Cheaha Wilderness of 710 acres that the Forest Service had requested in order to create a more manageable boundary there. Bill Nichols had readily assented and cosponsored the House legislation with Ronnie Flippo.

The legislation created Alabama's first and only national wild and scenic river, encompassing some fifty-two river miles of the West Fork Sipsey and its upper tributaries. All stream segments within the wilderness were given a "wild" designation, the national system's most protective classification, as were Thompson Creek and its two upper tributaries, Tedford and Mattox. Flannagin and Borden Creeks were given a "scenic" classification, which permits limited development and some access by roads. Montgomery Creek was restored to the bill (having been deleted by Heflin in his 1986 legislation as the site of his $20 million lake) and given "scenic" status. The river itself, as it flows southward out of the wilderness, was divided into "wild" and "scenic" segments.[9]

As Joe Brown had urged, and ultimately the Forest Service insisted, all roads within the expanded wilderness were closed, including the Northwest Road, and the Forest Service was directed to convert them into hiking and horse trails. There was one concession to motorized access, however. The eastern boundary of the wilderness was drawn so as to "lasso," and thereby exclude, about three miles of the Bunyan Hill Road as it runs westward from the Cranal Road down to Borden Creek. Birmingham Audubon Society stalwart Robert R. Reid Jr. was the source of this idea. Bob noted that this was one of the few places that a road offered entrance to the interior of an especially pretty Sipsey canyon for "old folks like me." Because Bunyan Hill Road had historically accessed a canoeing put-in site for the floatable portion of Borden Creek, the Forest Service agreed.

Instead of the $20 million dam and recreation site, the Forest Service was directed to "determine the feasibility" of such a lake in the northeastern part of the Bankhead Forest. At this writing, no such feasibility study has been performed.

The Forest Service was also directed to monitor the quality of water "flowing towards and into Lewis Smith Lake" and to take action "to prevent any national forest management activities from causing injurious water quality"—a restatement, in essence, of existing Forest Service policy.

And then, of course, there was the pine beetle. I was tempted to set out this section in full and let the reader divine its convoluted provisions unassisted, so as to get a taste for how many man-hours had to be devoted to its creation, but that would be too cruel. One can get the flavor of the thing from a summary.

First, the section restated the Secretary of Agriculture's existing authority under the Wilderness Act to control fire, disease, and insects, but now specifically including the southern pine beetle, within the expanded wilderness and the wild and scenic river corridor, on condition that the secretary first make a determination that there is a threat to adjacent lands, public or private, that lie outside the designated areas. Ironically, this last proviso actually strengthened the wilderness protection against logging to "treat" pine beetles.

Second, within five years, the secretary was directed to evaluate "the dangers and potential dangers of the Southern Pine Beetle and other insects in the Bankhead National Forest" and adjacent lands. If that evaluation produced a new decision to review Forest Service policy on the pine beetle, then the review was to include the wilderness and wild and scenic river. Thereafter, the secretary "shall monitor the dangers and potential dangers of such insects and such policy, closely."

At this writing, the Forest Service had performed no review of its southern pine beetle policy, and its new generation of forest plan for Alabama, released in 2003, provides for an entirely different way of addressing the problem, rendering the debates of the 1980s moot.

A couple of other sections of the act need to be noted. First, the Sipsey additions were brought into the national system under the original 1964 Wilderness Act so that no condemnation authority was granted to the Forest Service.

Second, there was something called "release language," which was to impact the effort to establish the Dugger Mountain Wilderness Area in the 1990s.

In the early days of the national RARE II legislative campaign, congressional Republicans and the Reagan administration sought, as a condition to any legislation designating a particular state's RARE II areas as wilderness, that all other national forest land within that state be "released" by the legislation to general multiple-use management and that the Forest Service be under no obligation to ever consider released lands for potential wilderness designation again. This was known as "hard" release language, and it created quite a stir when first proposed.

Because the Republicans controlled the Senate and the White House during this time, national conservation groups knew they had to make a concession on this issue, and compromise, or "soft" release, language was eventually agreed upon. It provided that the Forest Service would not be required to consider undesignated land for wilderness during the life of the forest plans then being developed in each state, a period of ten to fifteen years. And there would never again be a RARE evaluation in the affected state "unless expressly authorized by Congress." As had been the case with all the other RARE II bills that preceded it, the 1988 Sipsey Act contained this compromise language.

The last links in the chain of protection for the Alabama RARE II areas lay in the settlement of the Alabama forest plan appeal that David Carr and

SELC concluded with Joe Brown. As had been specified in the 1985 gentlemen's agreement negotiated by Mike Leonard, the Bankhead RARE II areas in the West Fork Sipsey watershed were classified as semiprimitive, but now Dugger Mountain would also receive that administrative designation, as would the Blue Mountain area north of Cheaha State Park. David and Joe also agreed that the "primary harvest method" for timber within the semiprimitive areas would be selective cutting, and that any clear-cutting would be only for "wildlife or recreation goals" and would "not exceed 10 acres in size."

Several other beneficial land management details were resolved by the appeal, which was dismissed by agreement in the spring of 1989, but the above were the most significant as far as the wilderness issue was concerned. We had achieved not just a doubling in size of the Sipsey Wilderness, but a vastly improved level of protection for the unique West Fork Sipsey watershed through the mix of wilderness, wild and scenic river, and semiprimitive designations there. And because those Bankhead semiprimitive areas still retain much of their wild character, they remain candidates for future additions to the Sipsey Wilderness, if ever the public will and the political climate again converge. In truth, when we began our participation in the RARE II process back in the 1970s, such a result would have exceeded our fondest hopes.

Even Charles Borden, who had advocated at one point in the campaign that the entire Bankhead Forest be designated wilderness, was pleased: "This is really a tribute to democracy in action . . . a small group of citizens working together for such a noble cause, who found a sympathetic ear in an elected official, Ronnie Flippo, who saw it through to fruition."[10]

[Ronnie Flippo said,] "at a time in which we constantly hear of environmental tragedies across the county, we as Alabamians should be proud that we have tucked away a small but beautiful part of Alabama for future generations to enjoy."

[Flippo] described his work on the passage of this bill as the "most gratifying thing I have done in my entire congressional career."[11]

"With all due respect to Alabama's logging industry," wrote John Brinkley of the *Birmingham Post-Herald*, "it appears that Alabama's environmentalists have finally won one."[12]

CHAPTER TWENTY-THREE

In Retrospect

THE TWO CAMPAIGNS TO ESTABLISH and enlarge the Sipsey Wilderness, and to create the West Fork Sipsey National Wild and Scenic River, resulted in one of the South's finest and largest areas of public lands under permanent statutory protection, and the second largest national forest wilderness east of the Mississippi River.

Ron Tipton, now senior vice president with the National Parks Conservation Association in Washington, D.C., says of the effort, "I think it's remarkable that we actually got a block of wilderness and wild and scenic river of that size in a state like Alabama. It transcends anything we [the Wilderness Society] were able to do in any of the other Gulf States. I was always intrigued with the Sipsey as an issue because of the magnitude of what we were trying to do there. It stands out in terms of my career in wilderness preservation."

If this book teaches nothing else, it is that statutory wilderness protection is first and last, beginning to end, a political process. Lawsuits, forest plan appeals, publicity campaigns, petition drives, and their ilk can prove fruitless if one or more politicians are not willing to put their prestige and political capital on the line to see a campaign to completion. In this light, Ronnie G. Flippo stands apart from all the other public officials who have supported Alabama wilderness. No other had to bear the humiliation and frustration that Ronnie Flippo endured or the disrespect exhibited by his counterpart in the U.S. Senate, who was not only a member of his own political party but also a resident of his congressional district. Overcoming this, Ronnie Flippo is more than justified in proclaiming the Sipsey expansion to be "a very proud achievement . . . the fondest recollection that I have from serving in the Congress [and] the most enduring legislation that I was associated with."

The former congressman now heads R. G. Flippo and Associates, a private lobbying firm in the nation's capital. He credits Charles Borden and two friends in Florence, Donald and Dee Patterson who were active with the

Shoals Area Audubon Society, with first calling his attention to the Sipsey and the enlargement opportunity that RARE II presented. "I went out and looked at it," says Flippo. "It's such a pristine and magnificent place that I got very interested in it—[it's] absolutely irreplaceable."

Of his long struggle with Howell Heflin, Flippo says, "I decided to keep the issue before him in the hopes that at some point he would see the desirability of passing [a wilderness bill]. Normally, when a bill is introduced and dies, you don't try again, but I decided that I didn't know of any significant reason why the bill shouldn't pass, so I decided to keep it before him and give him an opportunity to pass it."

The good senator has had occasion to reflect upon his treatment of Flippo during those years. In their biography of Heflin, John Hayman and Clara Ruth Hayman quote him as saying, "I have great admiration for Congressman Ronnie Flippo, and he was pushing it. I felt terrible not being able to support what he was trying to push through. Ronnie is a friend, and he had helped us with the Judicial Article when he was in the legislature. I finally went along as a favor to Flippo."[1]

When I interviewed Flippo for this book, I read this passage to him and asked for his comments. This was our exchange:

Flippo: All of my association with Howell Heflin has been in a most positive fashion. This was just a situation where he had a different belief. I guess it shows that people can disagree and still be friends. I still consider Howell Heflin an enormous friend and all my associations with him have been with positive results to me. I can only say that I have the highest regard for the senator.
Randolph: Well, Ronnie, you're very diplomatic and that's what makes you such a good politician and lobbyist, but let me give you my perspective on it. The man behaved as though you weren't there for six years and had nothing to do with it, and he didn't give any credence to what your desires were.
Flippo: [laughing] Well, I can't debate your point.

It seems only fair, then, to give Howell Heflin a last opportunity to justify his hostility toward Flippo's legislation. Quoting again from the Haymans' book: "I've got problems philosophically with just creating a wilderness. The Forest Service has done a pretty good job of timber management. You can't even build a road through a wilderness. . . . I'm not a wilderness fan. Oh, it's all right to have some of it, but if you've already got something like 20,000 acres, that's enough to wander around in and watch birds."[2]

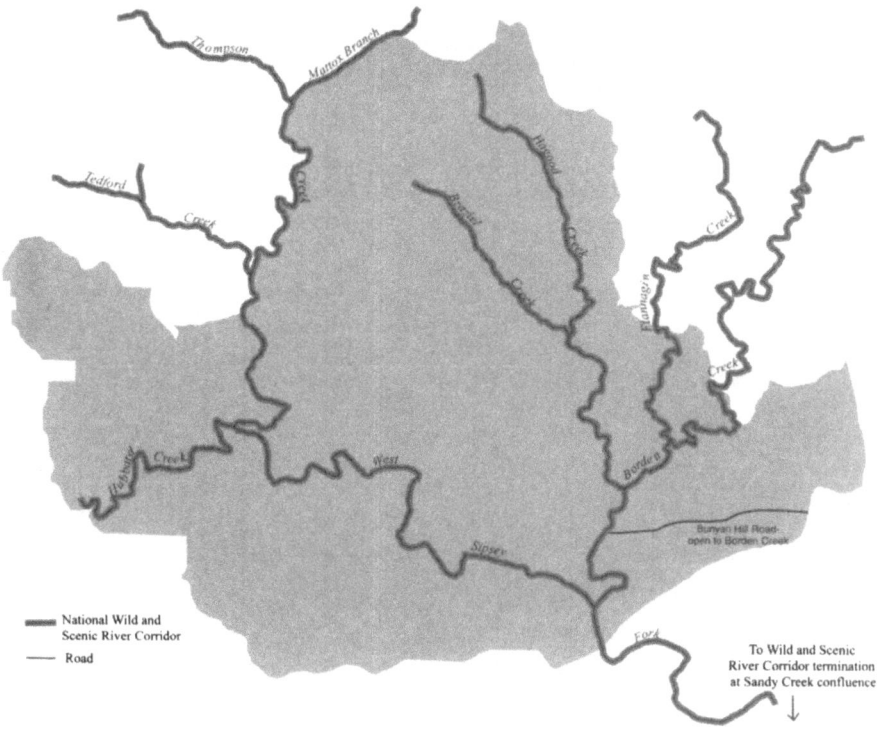

Expanded Sipsey Wilderness Area and West Fork Sipsey National Wild and Scenic River, Established 1988

PART FOUR

The Dugger Mountain Wilderness Area

Dugger Mountain Wilderness Area, Established 1999

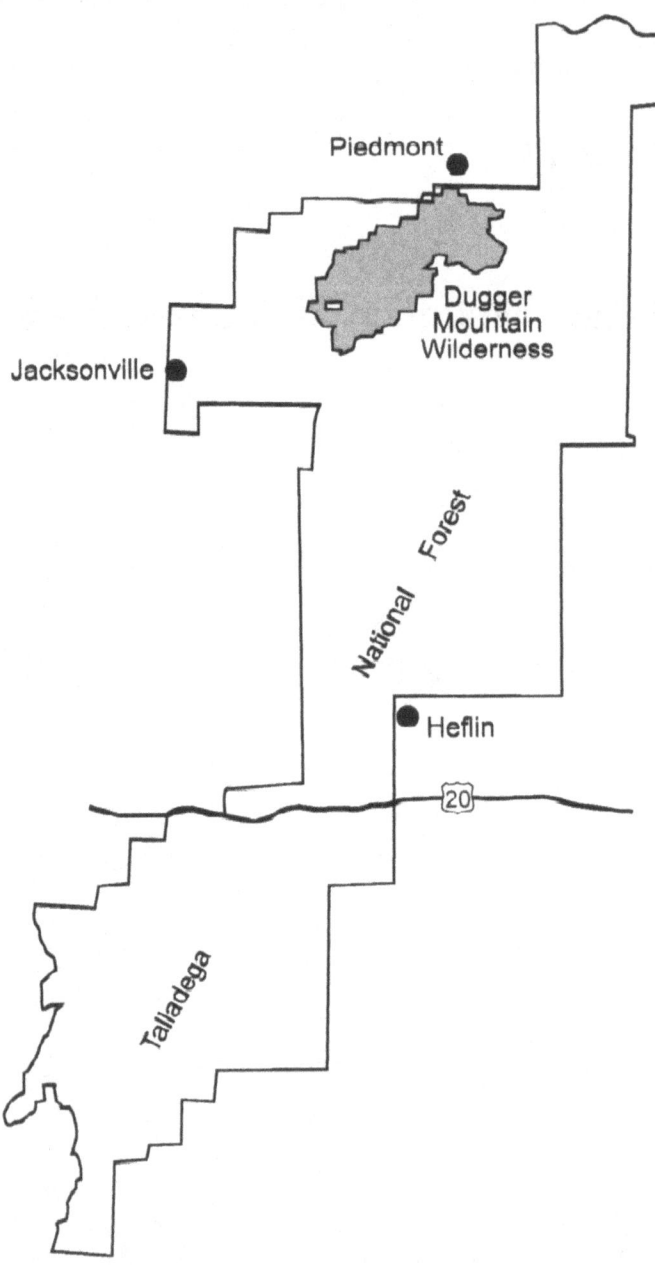

Dugger Mountain Wilderness Area, Established 1999

CHAPTER TWENTY-FOUR

Fire on the Mountain

BESIDES BEING ALABAMA'S "newest" designated wilderness, Dugger Mountain is also its least known, so a measure of introduction is in order.

Dugger lies at the northernmost extreme of the Talladega Mountain range.[1] Its crest, at some 2,140 feet, is Alabama's second highest peak, Cheaha Mountain being the highest. The town of Piedmont nestles at the northwestern base of Dugger. Only in that direction does one have a view of civilization. To the south and east lie the depths of the Talladega National Forest, and to the west, the heavily wooded Choccolocco Mountain range, the site of the recently created 9,000-acre Mountain Long Leaf National Wildlife Refuge. The surrounding foothills in these directions are also heavily forested, hiding both the sight and sound of civilization in the intervening valleys. Standing on Dugger, one gets the impression of being in a far bigger wild, where one can "see" primitive mountain Alabama much as the Native Americans saw it, and as very few have seen it in modern times.

To the layperson Dugger Mountain's geological place in the scheme of things may seem surprisingly clouded. Some twentieth-century geologists maintained that it was an extension of the Blue Ridge. Today, the weight of authority places it in either the Piedmont formation, or the Ridge and Valley Province, or perhaps both, with cherts from the Cumberland Plateau to boot. Whatever the correct classification, seen from the air, Dugger Mountain very much resembles peaks in North Carolina, with towering forested ridgelines over steep mountain ravines and large rock outcroppings along the summit. Dugger's ridges are a quartzite of the Weisner formation (for those interested in such things), while the northern slopes are underlain with limestone, giving birth to several springs. The southern slopes are of an "impervious Talladega Slate." Streams draining the mountain include headwaters of Choccolocco Creek and Terrapin Creek, where the town of Piedmont gets its water supply.[2]

During the 1990s, as a part of its wilderness campaign for Dugger Mountain, the Alabama Conservancy, now calling itself the Alabama Environmental Council, commissioned several studies of the natural history of the area by faculty and students of nearby Jacksonville State University. One of the studies was something called an "archaeological reconnaissance," which amounts to a quick field pass-through of a limited area. Despite the restricted nature of the reconnaissance, the team located two prehistoric rock shelters, one of which had been looted, but the other still containing artifacts indicating it to be a habitation site. Twenty-five stone mounds were determined to be "linked to prehistoric burial practices." The group also visited five other sites that had been previously identified by team members in several prior "archaeological pedestrian surveys." The team concluded that Dugger Mountain had "been continuously occupied by aboriginal groups from the Paleo-Indian/Pleistocene Period (10,000 B.C. to 7,000 B.C.) up to the 19th Century."[3]

Over time, the "aboriginal groups" evolved into the Creeks and Cherokees, and a portion of the traditional boundary between their nations lay somewhere in the vicinity of Dugger Mountain. Even the Indians were uncertain of the precise location, making war upon each other from time to time in retribution for perceived encroachments. With the defeat of the Creeks by Gen. Andrew Jackson in 1814, the U.S. Department of War appointed three commissioners to formally ascertain and survey a permanent boundary. According to the U.S. Forest Service,

> After several false starts and setbacks the line was finally surveyed by Gen. John Coffee and ratified by the president in 1830. The line still exists on some current maps, and passes between Dugger Mountain and Piedmont. Pursuing a rather erratic course, it crosses the northeastern portion of the Dugger Mountain Unit and passes into Georgia. It is interesting to note that the years of preparation and actual surveying of this line were virtually wasted since the line served no useful purpose and ceased to be of importance when the Cherokee were removed to western lands a few years later.

Following the Cherokee removal, white settlers entered the Choccolocco and Terrapin Creek valleys, and by the 1840s, a town known as Cross Plains existed at the site of present-day Piedmont. The mountain at that time was known alternately as Ladiga Mountain, after an Indian chieftain of that name, or Terrapin Mountain. It became known as Dugger Mountain after

Taylor Dugger, a one-legged veteran of the Civil War, established a farm at the northwestern base of the mountain and purchased the land running up the mountainside, which he managed as a wildlife refuge.

By the turn of the twentieth century, Cross Plains had become Piedmont and was served by at least two railroads, the Seaboard Air Line Railroad and the Selma, Rome, and Dalton Railroad. Encouraged by the railroads, private interests moved to take advantage of the era's widely held belief in the healing properties of mountain spring water. Two health resorts were developed in the Piedmont area. One was at the base of Dugger Mountain itself, known as the Borden–Wheeler Springs Resort. It featured a hundred-room hotel and "a village of cottages." The facility burned in 1935 and was never rebuilt. The other resort, the Piedmont Springs Hotel, was situated at an elevation of 1,200 feet on Blue Ridge Mountain southwest of town. It was abandoned in 1949.

One of Dugger Mountain's particular stories is that of Ferguson's Light, which dates from the turn of the twentieth century. It seems that one Stephen Ferguson of Piedmont developed acute allergies during the fall of each year. His doctor suggested that Stephen seek a higher elevation when symptoms struck. He chose Dugger, where each evening he lit a signal fire on the mountain to let his family in the valley know that he was still kicking.

In 1928 the U.S. Forest Service began purchasing private property centered around scattered tracts of land that were still in the public domain, meaning they had been seized from the Indians and never sold to private owners. The largest private landholder to benefit from government acquisition was a speculator named Wellington R. Burt, who had bought up hundreds of small farms and forest tracts on the mountain in hopes of a profitable resale.

In the modern era, perhaps no one personified the Dugger Mountain old-timer better than Pink Edward Burns of Rabbittown in Calhoun County. In the early part of the twentieth century, Pinky Burns's parents took up occupancy of an old one-room log structure near Rabbittown that had been built as a schoolhouse for the children of miners working a claim on a Dugger ridge known as Red Mountain. The landowner allowed the Burnses to enlarge the building to accommodate their family, which included Pinky and a sister. Following the death of his parents and the marriage of his sister, Pinky Burns became sole occupant of the old log house. Eventually, the U.S. Forest Service acquired title to the real estate and honored an "at will" tenancy agreement that the prior owner had granted the Burns family. Dugger Mountain was Pinky's lifelong domain.[4]

One day, so the story goes, Pinky Burns was visited by a man who said he was dying and had come to visit his ancestral homeland. The caller was a Cherokee Indian named Little Beaver, whose grandparents had fled to Dugger Mountain in the 1830s to escape the forced Indian removal then under way. They occupied a rock shelter on the mountain for a number of months, until compelled by hunger and the elements to seek refuge among Creeks and Seminoles in Florida. There, Little Beaver was raised, hearing stories of the great "rock house" in Alabama that had saved his ancestors from the Trail of Tears. Now Little Beaver asked Pinky to lead him to the shelter, where they camped for several days, performing ceremonies and sharing tales and companionship. Pinky never saw Little Beaver again, and died himself on November 24, 1999, at the age of eighty-one.[5]

Following acquisition by the federal government in the 1920s and '30s, Talladega National Forest lands entered into a period of reforestation and custodial management, which lasted into the 1960s. During this time Dugger Mountain's native hardwoods and longleaf and Virginia pines were encouraged by the Forest Service to regenerate. A policy of active fire suppression during the custodial period resulted in the thick stands of native pines now seen on Dugger Mountain's ridgelines. Isolated clusters of "relic" longleafs that have never been logged survive to this day. In fact, primarily because of general inaccessibility, commercial logging on Dugger Mountain was limited even during the 1970s and '80s, when production on national forest lands rose so dramatically. Selective cutting was a common practice, and only some isolated stands totaling a few hundred acres were actually clear-cut and converted to loblolly pine.[6] "Overall, it is not our best timber growing land," Cheryl Herbster of the Shoal Creek ranger district told the press.[7]

After the Forest Service hatched its plan for the Talladega Scenic Drive in the late 1950s, the agency began to see Dugger Mountain as something more than just a source of commercial timber. The scenic drive, in fact, was to be routed up its crest as a part of an overall plan to open the Talladega Mountains to "intensive recreation development."

In February 1973 the agency held a public hearing at Jacksonville State University to receive input on future management of what was then called the "Dugger Mountain planning unit," comprised of some 16,500 acres. The statements of participants reflected the growing public desire to have logging restricted, if not exempted, on the mountain. Some saw it as a cash cow in a different respect:

> [An] individual, representing Boy Scouts in the area, said his group was interested in seeing a hiking trail developed along the mountains in the Talladega Forest. "We have many scouts that want to get their 50-mile hiking badges but there is no trail in the state that is anywhere near that long," the spokesman said.
>
> Bruce Barton, representing the Tallacoosa Highland Lakes Association, told the audience [that] planning of the Dugger Mountain Unit is a step toward the development of a model area. Barton added that construction of the Scenic Drive would enable tourists from all over the South to enjoy the area and the available recreation.
>
> The president of The Alabama Conservancy, Mrs. Linsey Smith [sic] of Birmingham, said because of the high ridges and low creek beds a road through the area would destroy the watershed and eventually block the water drainage in the area. She also suggested that all motor vehicles be prohibited from the Dugger Mountain Unit.
>
> Piedmont Mayor L. H. Gunter said his concern with the development of the area lay in providing all the recreation facilities possible for tourists and campers. "We would like to see golf courses, swimming pools, swings and even horse shoe pitching rings," Gunter said. He mentioned the possibility of having a tourist information booth built near the entrance to the park.
>
> [Only] a spokesman representing . . . lumber manufacturers in the Southeast said the harvesting of timber must continue. "We are asking that the U.S. Forest Service continue under the multiple use program. That the forests be restricted to any one use would not be beneficial to the nation."[8]

Interestingly, even though the Forest Service was denying at the time that wilderness existed in the eastern United States, a "data summary report" released by the agency in 1974 in connection with planning for the Dugger Mountain unit contains an oblique reference to possible wilderness designation for the mountain. In a section on wildlife, the report observes that "Wilderness classification of the Unit, as compared to its present condition, would have little measurable effect on present wildlife populations." That was the only mention of the word "wilderness" in the document, but it went on to say, "A good portion of this Unit is relatively secluded and offers excellent opportunity for a variety of dispersed recreation activities, such as: hunting, hiking, primitive camping, picnicking, nature study, horseback riding, etc." The report concluded that "[i]n the future, more intense timber management will increase the volume and value of harvested timber products, but the biggest opportunity for economic enhancement is in the development of the recreation resource, which would increase employment and encourage tourism."[9]

Not until the Carter administration instituted its RARE II initiative in the late 1970s was Dugger Mountain officially touted by the Forest Service as a wilderness candidate. Early in the RARE II process, the agency conducted workshops in which citizens and forest rangers gathered at tables to "rate" the wilderness attributes of the various candidate areas. Because my personal project at the time, the Cheaha Wilderness, was in the same Talladega Forest Division as Dugger Mountain, I participated in rating both of the areas. The rangers were noticeably enthusiastic about Dugger Mountain—perhaps in part because their supervisor's office in Montgomery was then pushing Dugger as an alternative to Cheaha, but enthusiastic nonetheless. Following their lead, I joined them in rating Dugger Mountain as having some of the highest wilderness attributes in Alabama, an interesting learning process for me, as I had never visited Dugger.

Despite endorsement by the district rangers and a high wilderness rating, Dugger Mountain was "allocated to non-wilderness" by the Forest Service at the conclusion of RARE II, since the Alabama Wilderness Coalition proved to be unwilling to abandon its Cheaha proposal in favor of Dugger.

Dugger Mountain was next mentioned as a candidate for special designation in 1984. In September of that year the National Park Service released a report describing areas in the Appalachian Ranges having potential for classification as national natural landmarks, an administrative designation that "fosters protection of scenic beauty" and provides the "manager of a natural area with the incentive to carry out land-use practices which result in preservation of its natural qualities." This report found that Dugger Mountain "appears to be nationally significant" and advocated further consideration.[10]

Ironically, the next proponents of Dugger preservation to emerge were Alabama logging interests that were opposing the enlargement of the Sipsey Wilderness Area in the 1980s. During negotiating sessions sponsored by U.S. senator Jeremiah Denton in 1985, the Alabama Forestry Association, the Society for the Wise Use of Federal Forest Lands, and the U.S. Forest Service came forth with a proposal for an 8,400-acre Dugger Mountain Wilderness Area. Although these logging interests were motivated by a desire to severely restrict a Sipsey Wilderness enlargement, offering Dugger in its stead, the Alabama Wilderness Coalition eventually endorsed the idea as a means of inducing an agreement with the timber industry, thereby finally compelling Sen.

Howell Heflin to deal with Sipsey Wilderness legislation. Unfortunately, the U.S. congressman for the Dugger area, Bill Nichols of Sylacauga, opposed the new wilderness, writing to Sens. Howell Heflin and Jeremiah Denton:

> After considerable discussion regarding the proposal to place the Dugger Mountain area of the Talladega National Forest into wilderness designation and after having visited the area, I have come to the conclusion that I could not support legislation which would designate Dugger Mountain as a wilderness area.
> As you know, we established the Cheaha Wilderness area atop Cheaha Mountain some 25 miles south of the Dugger Mountain area, and I—and the locally elected officials in Cleburne County—have questioned the need for the Dugger Mountain Wilderness area which is so nearby.
> In light of the fact that I find little or no support from locally elected officials or from citizens themselves who live in Cleburne County, I have come to the conclusion that Dugger Mountain should be omitted from any wilderness legislation which you may be proposing in the United States Senate.[11]

Just prior to Nichols's decision, the Forest Service released its long-awaited 1986 Land and Resource Management Plan for Alabama. Since different legislative proposals for a Sipsey Wilderness expansion and a possible Dugger Wilderness were being kicked around at the time, the Forest Plan included a proposed 8,460-acre Dugger Mountain Wilderness within a "range" of some 19,000 acres of wilderness, including a Sipsey enlargement, which was acceptable to the agency. All the areas within this "range . . . will be protected to maintain current characteristics until Congress determines which areas will be Wilderness."[12]

When the Sipsey expansion legislation finally passed Congress in 1988, it contained something commonly referred to as "release and sufficiency language." These were provisions negotiated at the national level between conservation groups and congressional Republicans, with the agreement that they be included in all RARE II bills that passed the Congress. They included a declaration that the Forest Service's 1979 RARE II study was legally "sufficient" with respect to consideration of areas within the affected state for wilderness designation. They also provided that the Forest Service "would not be required to review the wilderness option" for "old" RARE II areas in the state's forest plan that was then being finalized for the coming ten to fifteen years. The effect, then, was to "release" Dugger Mountain, as far as the Forest

Service was concerned, from wilderness consideration until the next Alabama forest plan was developed.[13]

Nevertheless, in point of fact, Dugger Mountain remained a wilderness-in-waiting—waiting, primarily, for new blood, new grass-roots leadership. We weary wilderness warriors of the RARE II era had done all we could. It was time for us to fade away, to vacate the stage in favor of a new generation of citizen activists who would take up the Dugger Mountain cause and see it through to a successful conclusion.

CHAPTER TWENTY-FIVE

Songs of the Summit

ANY INTRODUCTION OF DUGGER MOUNTAIN activists must begin with one W. Peter Conroy.[1] A native of Scranton, Pennsylvania, Pete spent his adolescence in Asheville, North Carolina, where he was immediately struck with what he calls the "western North Carolina scene." His instant love for the mountain wilds motivated him to begin a lifetime of environmental advocacy. Graduating from Furman University with a degree in biology, he moved on to the University of Georgia, where he obtained a master's in zoology and began to focus his career toward environmental education.

An opening for a curator's position at the Anniston Museum of Natural History brought Pete Conroy to Alabama in 1985. The mountains of northeast Alabama reminded him so much of Asheville that he resolved to stay. Soon his natural leadership abilities would send him to the head of Alabama's environmental community and embroil him in the wonderful world of Alabama politics.

In their affinity for politicians and the political game, Pete Conroy and Mike Leonard are very much alike. They are, in fact, good friends. But whereas Mike generally prefers to direct his energies behind the scenes, Pete seems to attract political appointment and visible leadership roles. Pete's résumé is so replete with volunteer and public positions that one wonders how he has had time to earn a living. A few of the highlights include service in Gov. Jim Folsom Jr.'s executive cabinet as liaison for environmental affairs, appointment by Pres. Bill Clinton as alternate federal commissioner of the ACT/ACF (Alabama-Coosa-Tallapoosa/Apalachicola-Chattahoochee-Flint) River Basin Commissions, appointment by Gov. Fob James to the board of directors of Alabama's Forever Wild program, and appointment by Gov. Don Siegelman as cochair of the Alabama Millennium Trails Committee. Since 1995 Conroy has been employed as director of Jacksonville State University's Environmental Policy and Information Center.

During his years with the Museum of Natural History, Pete became acquainted with Anniston attorney (and later state senator) Doug Ghee, who encouraged Conroy to channel his energies into creating a conservation group that might benefit the local area. Literally "shopping" the various existing environmental organizations, Pete settled upon the Alabama Conservancy and, with other local activists, "pulled together" an Anniston area chapter of the group. From there, Pete progressed to the Conservancy's state board of directors, and, in 1989, to its presidency, a post he held for nearly five years.

Among the federal projects Pete worked on during this time was Mike Leonard's effort to connect the Pinhoti Trail to the Appalachian Trail, something that occasionally required Congress to appropriate funds for corridor acquisition. Thus, Conroy found himself often in the offices of Alabama's congressional delegation. He became friends with the likes of Sens. Howell Heflin and Richard Shelby—not, Pete says, by talking politics with them, but by avoiding politics and sharing other mutual interests, such as hobbies.

"I learned with Judge Heflin," says Conroy, "to never ask him direct questions. He just might give you the wrong answer. I'd say, hypothetically, if such-and-such happened and so-and-so would be willing to support it, do you think this or that could be done. And then he would give me a qualified probably." With Senator Shelby, "we'd talk about his passion for collecting Persian rugs and pie recipes. He said I was the only conservationist in Alabama that he could stomach."

The new Anniston chapter attracted the other key players in the Dugger Mountain story, Francine and Bruce Hutchinson.[2] Bruce was retired from the Anniston Army Depot, where Francine was employed. They were serious people who, like Pete Conroy, wanted to work on meaningful projects. Soon, the three of them were asked to join the board of directors of the Alabama Conservancy, and became the triumvirate that would lead the Dugger Mountain campaign in the years ahead.

Francine credits Mary Burks with suggesting that the preservation of Dugger Mountain was the kind of serious project they were seeking. The idea immediately "resonated" with the Anniston Conservancy chapter, Pete Conroy says. Early on, "we considered different alternatives for protecting the mountain. Would this be a scenic area, a recreational area? We ran through all the different designations that would allow for varying levels of management, but we kept coming back to wilderness."

In 1992 the Conservancy board of directors formally approved a Dugger Mountain Wilderness campaign and eventually allocated funds to support it.

It happened, as well, that Francine Hutchinson had the opportunity to attend graduate school at Jacksonville State University, and to further the wilderness effort, she decided to devote her master's thesis to the flora of Dugger Mountain, something that would lead to a more thorough examination of the area's natural history.

From the beginning Francine was drawn to the mountain: "[T]he songs on the Dugger summit are truly unique. The wind blows from any and all directions to converge in pines and broadleafs, whispering secrets, and telling thankfulness. The wind winds around and in-between trees, craggy granite outcrops, and deflects off the steep spine of the mountain itself. I have been to some of the world's most incredible places, and nowhere else have I heard a song like this."[3]

Encouraged and mentored by her professor, David Whetstone, Francine spent several years researching the natural history of Dugger Mountain. "It's got quite a few different habitat sites," she says. "It's got seeps; it's got bogs; it's got huge rock outcrops. From the Alabama upland perspective, it's got all the range of habitats, like a microcosm of the ecological region. The summit is at the border between the Piedmont and the Ridge and Valley Province, and where you've got different regions coming together like that, they are the richest ecologically."

She was also delighted to discover along the summit a growth of American chestnut trees, a species nearly driven to extinction by a blight introduced in the early twentieth century:

"One day we counted 50 specimens in a one-mile stretch . . . [including] a struggling 30-foot . . . tree. In the five years since we have been watching it, the tree has steadily weakened from the blight. But maybe this tree or another one like it somewhere on the mountain might be the breakthrough chestnut that has braved the genetic labyrinth to produce blight resistance. If given time, peace, and protection, the chestnut may return. Wilderness areas will be key to nature's hospital laboratory."[4]

To achieve wilderness designation for Dugger Mountain, the three leaders resolved to take a careful, well-thought-out approach. Pete Conroy says, "I'd studied how the Sipsey expansion went wrong and went right, in a postmortem. I kept thinking that we could avoid a lot of those problems by first getting strong local support."

Bruce Hutchinson puts it this way. "We found out, to get this thing passed, we had to get every politician in Calhoun and Cleburne Counties, every member of our legislative delegation in Montgomery, every county commissioner and mayor, all to be 100 percent behind this. We had the impression, rightly so I think, that one person could kill it."

Thus began what may arguably have been the single most important effort to garner support for the new project. Individually and together, Bruce Hutchinson and Conservancy activists Ken Wills and Hendrick Snow visited every private landowner around the base of the mountain and obtained their signatures on a petition for wilderness designation. The result was "virtually unanimous," ninety-two out of ninety-five owners, a powerful tool in convincing local politicians to support the proposition.

Of the three holdouts, Bruce says that one told him, "'I agree with what you're doing, but I belong to the timber industry, and my friends would shun me, so I'm not going to sign it.' Another guy said, 'I'm not going to sign it because I've been to the Forest Service and they told me that if you get Dugger Mountain a wilderness they're going to stop hunting. And don't even bother going over to my brother's house, because he's not going to sign it either.' I told him that what he said was wrong, but he wouldn't listen. So those were the three holdouts, but to my knowledge they never did anything to actively oppose [us]." The Forest Service's official position regarding the wilderness proposal was that they were required by the "release language" in the 1988 Sipsey legislation to defer consideration of a Dugger Mountain Wilderness until the next Alabama forest plan could be developed. Nevertheless, reported the *Birmingham News,* "Conservancy leaders say ... Forest Service officials have been helpful and appear to favor the wilderness plan."[5] However, says Bruce Hutchinson, "We discovered that it was going to take a long, long time to sell the Forest Service all the way to Washington."

Taking a page from the original Sipsey Wilderness campaign, Bruce and Francine organized a series of studies of the natural history of Dugger Mountain by local scientists to bolster the validity of the wilderness proposal. They obtained nominal funding from the Alabama Environmental Council (AEC), as the Alabama Conservancy was now calling itself. The funding "barely covered out-of-pocket expenses," says Francine. "Everyone did their work for 'the cause.'" The studies included:

"An Archaeological Reconnaissance of Selected Portions of Dugger Mountain, Calhoun County, Alabama," by Angela D. Morgan, Curtis E. Hill, and Harry O. Holstein of the Archaeological Resource Laboratory, Jacksonville State University

"A Preliminary Survey of the Mammals of Dugger Mountain," by Rick Wiedenmann of the Anniston Museum of Natural History

"Amphibian and Reptile Survey of Dugger Mountain," by Jason Adams, Eric Blackwell, George Cline, and Paul Rogers, Jacksonville State University

"Bird Survey of Dugger Mountain," by Mary Belue and George Cline, Jacksonville State University

"Water Quality Analyses of Nances and Terrapin Creeks," by Donald McGary and Michael K. Barnwell, Biology Department, Jacksonville State University

"Benthic Macroinvertebrate Bioassessment of Nances Creek, Terrapin Creek, Dry Creek, and Jones Branch on Dugger Mountain in Talladega National Forest," by Michael K. Barnwell and Frank A. Romano, Biology Department, Jacksonville State University

"Flora of Dugger Mountain," by Francine N. Hutchinson

"The Fishes of Dugger Mountain, Talladega National Forest, Alabama—An Icthyological Survey," by W. Mike Howell, Biology Department, Samford University, and Scott Linton, Deputy District Attorney, Jefferson County, Alabama.

These reports, combined with the 1974 "Data Summary, Dugger Mountain, Unit 1," by the U.S. Forest Service, comprised an effective feasibility study for the proposed wilderness, establishing its ecological significance and its place as "an island in the sea of history."[6]

All these efforts took not just months, but years. In the meantime, the Forest Service began implementing a series of management activities—herbiciding, occasional clear-cuts and roads to service them, prescribed burns, and the like—that Dugger supporters feared would damage the mountain's integrity.

Bruce Hutchinson observes, "I don't say that the Forest Service intentionally had a plan to kill the wilderness proposal, but some of their people had that in mind, because they came up with too many little things that they wanted to do, over and over again, that would detract from our ability to have a wilderness."

The problems with the agency reached a climax in early 1996, when the Forest Service came forward with several proposals to log portions of the mountain to "salvage" damage from Hurricane Opal and an outbreak of pine beetles, and to erect a communications antenna on an existing unused fire tower at the Dugger summit. The AEC's executive director, Pat Byington, complained to Regional Forester Robert Joslin, "In my 7 years as Executive Director at the Council, I have never seen such an onslaught of proposals and activities on such a small area. Especially an area which has seen virtually no activities for the past ten years. . . . My organization would like to propose a possible 'time out' on these activities in the form of a temporary moratorium. . . . My members are tired of having to respond to proposal after proposal."[7]

Fortunately, a friend from the Sipsey expansion era was active in Alabama at the time and was available to help. David Carr with the Southern Environmental Law Center (SELC), the Charlottesville, Virginia, public interest law firm founded by Birmingham native Rick Middleton, was already working on formal appeals of several Talladega National Forest timber sales. "It seemed like the Forest Service had it in for Dugger," says Carr. Acting for both the SELC and the AEC, David presented objections to the Dugger management activities, including a formal appeal of the pine beetle suppression plan. The net effect was that the hurricane salvage and communications tower plans were shelved by the Forest Service. Although the formal appeal of pine beetle suppression was denied, Carr says that a meeting with the Alabama forest supervisor and the regional forester "achieved a concession that they would [create] no new roads and were not likely to log unless private lands were definitely threatened."[8]

Perhaps, however, the most fruitful impact of these controversies was that the Forest Service was put on notice—had there ever been any doubt about it—that the Dugger Mountain wilderness campaign was in earnest. It also did not hurt that Congressman Glen Browder, who had been elected to the U.S. House of Representatives after Bill Nichols's death, announced during this time his intention to sponsor Dugger Mountain legislation. Pete Conroy says that Browder "was our first hard sale. He was never anything but positive about it." On September 17, 1996, Browder introduced H.R. 4087, the Dugger Mountain Wilderness Act of 1996, containing a total of some 14,000 acres that included not just Dugger Mountain but also a nearby ridge known as Oakey Mountain.

Today, Browder plays down the significance of his support. By the time he introduced his bill, he had become a "lame duck" congressman after an unsuccessful run for the Democratic nomination for U.S. Senate. The second session of the 104th Congress was nearly over, and the legislation had no likelihood of passage. He credits Pete Conroy with any contribution his legislation may have made. "Really," says Browder, "I can't claim anything other than being there procedurally and providing the assistance of my office as Pete pushed the movement for Dugger."[9]

I beg to differ. The bill got the Forest Service's attention, and its size—14,000 acres, which extended beyond Dugger Mountain and enveloped two public roads—got the agency to begin thinking about just how big a Dugger Mountain Wilderness ought to be. After all, it had gone from a RARE II "inventory" of 4,900 acres to an 8,400-acre proposal in the 1986 forest plan to Browder's 14,000 acres. Workable boundaries needed to be devised.

The legislation also provided a precedent for future congressional sponsors to point to, and it represented a 180-degree change in attitude from that of Browder's predecessor, Rep. Bill Nichols. Ironically, Nichols has been given credit in certain press reports for supporting Dugger preservation. Some even claim, erroneously, that he sponsored wilderness legislation for the area.[10] The truth is that Nichols prevented Dugger from being joined with the Sipsey Wilderness expansion legislation of the 1980s, and even went so far as to file what he called a "notice of appeal" when the Forest Service included a potential 8,400-acre Dugger Mountain Wilderness in its 1986 forest plan.[11] Glen Browder deserves credit for advancing the Dugger effort. Bill Nichols deserves none at all.

Moreover, any regrets one might entertain that Browder did not act sooner in his congressional career quickly collide with the reality that Howell Heflin, no friend of wilderness, was still in the U.S. Senate at the time. As it developed, the Alabama Forestry Association attempted to stall the Dugger Mountain proposal when next it came before Congress. Had Heflin still been in office, the outcome might have been quite different. Instead, Dugger Mountain benefited by the injection of new blood into Alabama's congressional delegation—and these folks were Republicans.

CHAPTER TWENTY-SIX

We Did This One Right

THE YEAR 1997 WAS A pivotal one, for it saw the installation of the future sponsors of the Dugger Mountain Wilderness in the U.S. Congress. Gone was Howell Heflin, bane of environmentalists, his Senate seat taken by Jefferson Beauregard "Jeff" Sessions III of Mobile. Also gone was third district congressman Glen Browder, and in his place stepped a future governor of Alabama, Robert "Bob" Riley of Ashland. Both men were Republicans.

Since Browder would not be returning to Congress, Dugger Mountain leaders concentrated on Bob Riley as soon as he began campaigning in 1996. Pete Conroy and Bruce Hutchinson met with him on several occasions, and every time he would appear in the area, they would have two or three people, local public officials when available, go up to him to express their support for a Dugger Mountain Wilderness. "Every time I saw him," says Bruce, "I reminded him about the project, that we would continue to ask for his support." Riley was positive from the start.

However, once Riley was elected to Congress, the Dugger leaders made a conscious decision not to ask him to sponsor legislation during his first term. Bruce Hutchinson says, "We didn't have all our ducks in a row. We were still working on the Forest Service and had some local public officials to convince. Second, we wanted to give him enough time to get his feet on the ground, learn who the players were and get some clout in Congress." Also, says Pete Conroy, "Riley was very busy dealing with the threats to [close] Fort McClellan and the plans for a chemical weapons incinerator at the Anniston Army Depot. We decided the time just wasn't right."

And then there was Mac Smith to be dealt with. Not to be confused with Scoutmaster Mac J. Smith of Montgomery, who played a key role in the Cheaha Wilderness effort, this Mac Smith once held the dual posts of Cleburne county probate judge and county commission president, and was one of the last holdouts among local officialdom. He had become convinced back in the 1980s that if Dugger Mountain were to be removed from the timbering base, Cleburne County would receive less revenue from the Forest Service in payments the agency made in lieu of property taxes. Bill Nichols had, in fact, used Smith's opposition as one of his reasons for keeping Dugger Mountain out of the 1988 wilderness bill. Now, Smith was telling Bruce Hutchinson, "I'm going to oppose this until the day I die, because it's going to economically impact my county." By now, however, there been a change in the way the Forest Service computed payments to the counties. The district ranger's office in Heflin was experiencing a gradual change in personnel and attitude, as well, and Bruce recruited one of the new officers, Cheryl Herbster, to convince Smith that he was mistaken. "That worked," says Bruce, "along with the support we were able to get from the newly elected probate judge, Monroe Lipscomb. It diffused the situation."

Changes in the Forest Service extended beyond the district rangers to the supervisor and his staff in Montgomery. "We sort of got away from the old timber beast that we used to have in the Forest Service," says Bruce Hutchinson. New people with broader educations and new attitudes were coming in, and they sought out the help of Pete Conroy and the Hutchinsons in developing the next generation of an Alabama national forest plan, which would be released in 2003. Dugger leaders used the opportunity of this cooperative effort to convince the agency to support the wilderness proposal and to help define its boundaries.

Thus, when Bob Riley was reelected to Congress in 1998, Dugger leaders felt the time had come to ask him to introduce wilderness legislation. Pete Conroy also decided it was time to approach the state's timber industry, hoping to prevent the kind of havoc that it had wrought upon the Sipsey Wilderness expansion proposal in the 1980s. Accordingly, in April 1999, Pete asked state senator Del Marsh to invite his hunting buddy, John McMillan, executive vice president of the Alabama Forestry Association, to a meeting to discuss Dugger Mountain. Pete says McMillan "was the first person I had encountered who said he didn't like this wilderness proposal. He said that his association was philosophically opposed to wilderness." After Pete detailed all the local support that had been generated, and Riley's promise to sponsor legislation, McMillan said the AFA might not oppose Dugger Mountain if it could be assured that outbreaks of fire and pine beetle infestation could be controlled so that land adjoining the wilderness was not threatened. Pete promised to research Forest Service regulations on these points and get back with McMillan.

Obviously the AFA promptly contacted Congressman Riley, for shortly after Conroy's meeting with McMillan, the press reported:

U.S. Representative Bob Riley said he plans to introduce legislation to make Dugger Mountain a Wilderness Area—as long as he can satisfy some concerns. . . .

He has two main concerns—the Southern Pine Beetle and fire protection. If the Southern Pine Beetle becomes a problem and threatens private property near the forest, Riley would like the owner to be allowed to control the infestation. Similarly, if there is a fire on a wilderness area that threatens privately owned property nearby, Riley wants firefighters to be able to properly control the fire.[1]

Shortly thereafter, David Carr with SELC helped Pete Conroy provide both McMillan and Riley with documentation establishing that control measures are permitted within wilderness where adjoining private land is threatened. Pete believed, then, that he had convinced the Alabama Forestry Association not to oppose a Dugger Mountain Wilderness bill.

However, it was apparent that someone from the timber industry had been attempting to sway the congressional delegation against Dugger legislation. Bruce Hutchinson says that when he and Pete Conroy traveled to Washington with representatives of the Alabama Chapter of the Sierra Club, the Southern Environmental Law Center (SELC) and the Southern Appalachian Forest Coalition (SAFC) to meet with Senator Jeff Sessions, "We learned that Sessions was being told that if we got Dugger Mountain it would significantly impact the timber industry. I said that the entire national forest is less than 3 percent of available timber in Alabama, and Dugger Mountain is less than 1 percent of the national forest, very insignificant. Sessions then said, 'I've got no problem with it whatsoever.'"

Actually, the senator was ahead of Bruce on this issue. According to Gerry Gilligan, his legislative assistant at the time, Sessions had already been looking at possible wilderness designation for Dugger Mountain, quite independently of anything Bob Riley was doing. Says Gilligan, "The senator was on the Environment Committee, and he had taken a strong stand on some controversial environmental issues, essentially painting him as an environmental bad guy, which really wasn't accurate. He was an outdoor person and had an appreciation for natural areas. He had been looking for something positive to do, so when Pete Conroy suggested Dugger Mountain to us, the senator focused on it as something to help his credibility on environmental issues. He knew that the national forests are not important sources of timber in Alabama."²

Knowing this compels me to credit Jeff Sessions with the first breath of common sense on the wilderness issue to come from Alabama's Senate delegation since the days of Jim Allen and John Sparkman. I wistfully imagine the outcome in 1982 had Howell Heflin and Jeremiah Denton simply acknowledged the economic insignificance of the Sipsey expansion proposal, respected the wishes of its House sponsor, and stood up to the timber industry. But the past cannot be changed; it is more productive now to simply underscore the enlightened contributions of Jeff Sessions—and of Congressman Bob Riley.

Meanwhile, the Dugger Mountain effort suddenly became more complicated. David Carr, Peter Kirby with the Wilderness Society, and other southeastern environmental leaders had created the Southern Appalachian Forest Coalition to encourage national forest roadless area preservation. David Carr says, "We got a letter signed by Senators Warner [VA], Thurmond [SC], Robb [VA], Hollings [SC], and Cleland [GA] calling on the Clinton administration to adopt a moratorium on roadbuilding in roadless areas in the Southeast." Through this effort, SAFC had established a relationship with Jeff Sessions and his staff, and had "almost got Sessions to sign on."

When it became apparent that there would be Alabama wilderness legislation, SAFC began to envision Dugger Mountain as the catalyst for a regional wilderness bill for the Southeast. Then, the prospect of Dugger Mountain legislation caused other wilderness advocates within Alabama to consider asking Sessions to sponsor a bill including a "wish list" of other areas in the state besides Dugger.

On May 6, 1999, David Carr met in Washington with Riley's and Sessions's legislative assistants. Riley joined them briefly, but Senator Sessions stayed to talk about the concept of a regional wilderness bill, including the wish list of Alabama areas.

> Looking over the maps, Sen. Sessions expressed particular interest in the Sipsey. He said he's always been amazed by the beauty of the forest and canyons on the Sipsey. . . . He said that he would be interested in helping to get other senators together [for a regional bill]. . . .
>
> Sessions noted that the timber industry is very big in Alabama. He said that they are concerned about more land getting locked up and being managed in a way that's not generating timber for the industry. . . . Sessions emphasized that it would be

best if we could work out some understanding with the timber industry and others whereby key areas that we wanted to protect got protected and other areas were definitely kept open for logging.[3]

Once Sessions and Riley left the meeting, their assistants suggested that SAFC and the folks in Alabama come up with a specific proposal, with maps, for their principals to consider. David Carr reports, "In our discussions immediately after the meeting, we agreed that the proposal should be broader than just the roadless areas. We should bring in some of the scenic areas that [the AEC] had proposed. In short, we should start bigger and be willing to let it get whittled back down."[4]

Wilderness advocates in Alabama moved to reach consensus on the wish list, an idea formulated as early as 1993 when leaders of the group known as Wild Alabama incorporated the Alabama Wilderness Alliance. The brainchild of Ned Mudd, Lamar Marshall, and Ray Vaughan, the new alliance was intended to advocate more wilderness designations statewide and to litigate in favor of natural area preservation, as well as support the burgeoning Dugger Mountain effort.

Thus, in July 1999, the alliance convened an in-state meeting to discuss a legislative approach. In addition to the Wild Alabama contingent, there were representatives of the Alabama Environmental Council, the Sierra Club, WildLaw, the Biodiversity Legal Foundation, Americans for Our Heritage and Recreation, and the Nature Conservancy of Alabama. David Carr came down from Virginia to represent the Southern Environmental Law Center.

Nearly 300,000 acres in Alabama's national forests were tentatively identified for possible special designation, with a Dugger Mountain National Wilderness Area as the centerpiece of the groups' proposal. A wish list posed the obvious problem that it would likely rile the timber industry more than would a single wilderness bill and might attract opposition from a variety of other sources as well. This sparked a discussion of whether the groups should entertain compromise, that is, the deletion of one or more proposed areas in order to obtain protection for others, or release language permitting undesignated general forest areas to be logged. The concept alarmed Ray Vaughan.

Vaughan had been a young assistant attorney general under Don Siegelman assigned to help with the Alabama forest plan appeal in the late 1980s. Now, he was a private attorney in Montgomery, directing a public interest law firm called WildLaw, which was making a name for itself challenging national forest management decisions and actions by government agencies detrimental to wild areas and endangered species. Vaughan considered compromise on natural area preservation to be a "sell out," and he wanted the groups at the Wilderness Alliance meeting to adopt and sign an Alabama Wilderness Protection Pledge that read as follows:

> I hereby pledge, promise and guarantee that I and the organization I represent will support only bills that will protect federal lands in Alabama as wilderness or as other protected areas and that do nothing else. I and my organization pledge to fight against any and all proposed bills that contain strings, riders or quid pro quo mandates. Such strings, riders and quid pro quo mandates include, but are not limited to, "hard release" language that mandates roadless or other areas not designated wilderness be put into the timber base, provisions that mandate logging or increased logging anywhere else, amendments or provisions that have nothing to do with protecting wilderness or other special areas, provisions that weaken the protections of the Wilderness Act of 1964 in any way for any area, and amendments that weaken any other statute or requirement of law.[5]

David Carr politely declined: "SELC agrees that the Dugger bill should be a clean bill with no strings attached. I feel confident that will be the case. As a result, I don't feel it is necessary to sign on to the pledge. . . . [T]here are just too many unknowns to say absolutely that we would not be willing to give up something, albeit very small, to get a whole lot more. Thus I think it may be premature to talk about absolutes. . . . The goal is never to give anything up but there can be situations where small concessions can secure big victories."[6]

Vaughan responded vigorously, arguing that if the Wilderness Coalition groups of the 1980s had adopted a pledge such as he was advocating and had not consented to "release language" in the Sipsey expansion legislation, then Dugger Mountain would have achieved wilderness designation along with the Sipsey in 1988.

However, this argument was founded at least in part upon a misconception that was then making the rounds: that Congressman Bill Nichols had supported a Dugger Mountain Wilderness Area back in the 1980s. In point of fact, however, Nichols actively opposed it, and without his support, there was no hope of getting a Dugger Mountain designation at that time. Nor would the Sipsey expansion legislation have passed the Republican-controlled Senate had it not contained the compromise "release language" negotiated at the national level for RARE II legislation.

Jeff deGraffenried, then president of the AEC, reacted to Vaughan's arguments in this way: "I am not comfortable with the language [Vaughan] is using at this time. I know we ALL want a clean bill. It may be a bit premature to start waging war (especially in the enviro community) so early."[7]

Dugger leaders Pete Conroy and Francine and Bruce Hutchinson were also uneasy with the way this was heading. "We didn't want to bog down Dugger Mountain with controversy over other areas," says Bruce, "so we asked Congressman Riley to sponsor Dugger alone." As a result, all the smoke and fury over compromise dissipated. The wish list also never materialized, at least not at that time, though it did eventually lead to a comprehensive proposal for public land preservation (see epilogue). On August 2, 1999, Rep. Bob Riley introduced H.R. 2632, the Dugger Mountain Wilderness Act of 1999. This, the fourth wilderness proposal for the mountain since the 1970s, was for 9,200 acres. The difference between H.R. 2632 and Rep. Glen Browder's 14,000-acre bill lay primarily in an adjustment in eastern and western boundaries to eliminate the two public roads that Browder had proposed to include. Oakey Mountain had been deleted, so that the eastern boundary was now Forest Road 500. To the west, the boundary was King's Gap Road, which crosses the southern flank of Dugger Mountain. Otherwise, the proposed wilderness contained virtually all national forest land lying between Alabama Highway 9 on the north and County Highway 55 on the south.

Riley told the press that,

his position as a Republican will give the legislation "a better shot than it had in the past. Generally, preservation legislation comes from the other side of the aisle," he said. "On our side there's a certain amount of hesitancy, especially from those from the west side of Alabama." It will be easier for a Republican to pass the legislation, he said, especially since the land has little value to any industry but tourism.

"Combined with the Chief Ladiga [Trail, an abandoned railroad bed now devoted to hiking and bicycling] and [the] Pinhoti . . . that area [Dugger Mountain] will likely start bringing in a tremendous amount of eco-tourism," Riley said. "And it's appropriate. The steep slope of the mountain makes it impractical for the timber industry, but it's quite scenic, quite beautiful."[8]

Unfortunately, the timber industry did not agree. Riley's office began getting letters from other areas of the state questioning the wisdom of creating another wilderness area. The letters "were probably prompted by a trade association," says Shana Jones, Riley's legislative assistant at the time. Although the congressman never received any formal opposing statement or letter from the Alabama Forestry Association, says Jones, John McMillan met with Riley on more than one occasion to contest the measure. Another AFA staffer, Boyd Kelly, confronted Shana Jones, and "we were certainly at odds over this."[9]

When Bruce Hutchinson heard about the AFA's intervention, he contacted Congressman Riley. "Leave John McMillan to me," said Riley.

Pete Conroy believes that the AFA's actions only stiffened Riley's resolve to proceed with the bill. Shana Jones agrees that Riley "can certainly be stubborn" but that the key to his actions was the simple fact that the Dugger Mountain Wilderness was overwhelmingly supported in his congressional district. "Congressman Riley truly represented his constituents," Jones says, "and this was something his constituents wanted. The Alabama Forestry Association got to the point where they realized they weren't going to change his mind and threw in the towel."

The AFA had a similar experience with Senator Sessions. His legislative assistant, Gerry Gilligan, says Sessions knew that

> [t]he profitability of pulling timber out of Dugger Mountain was a dubious proposition at best. So when John McMillan and Boyd Kelly came to talk to us, the senator was well prepared. He heard them out but told them he was going forward with it anyway. It was kind of a nervous moment for me. We were trying to do something good, and here were these people raising a bit of stink about it. But in the end, they did nothing. I think they understood that Senator Sessions would support them on really important issues and Dugger Mountain really wasn't worth alienating him over.

By October 1999, when Congressman Riley's legislation was due to be heard before the House Resources Subcommittee on Forests and Forest Health, Dugger supporters had achieved the distinct milestone of gaining "official" endorsement of the wilderness from the U.S. Forest Service. Alabama had a new national forest supervisor, James A. Gooder, who assumed his duties in September 1999, after Riley had introduced his legislation. "It was out with the bad and in with the Gooder," said Ray Vaughan.[10]

Bruce Hutchinson adds, "Although we had problems with the Forest Service in the early days of the Dugger campaign, I can't say enough about the support we received from them in the end, especially from Jim Gooder, [District Ranger] Earl Stewart, and other managers."

Congressman Riley and Senator Sessions asked the new forest supervisor to come to Washington with Pete Conroy to discuss his views on a Dugger

Mountain Wilderness. Gooder says that even though "there was a lot of impetus for Dugger moving forward, had I not been in agreement with it," there might have been a different result. Thus he made an immediate effort to travel the state to identify special areas that deserved attention: "It became pretty apparent to me that that whole Talladega Mountain spine, really everything from Rebecca Mountain all the way [north] to the Georgia line, is a special place, really spectacular. Dugger became a no-brainer for me. We had a lot of citizen involvement that said this was the right thing to do. The congressional delegation felt we could make it work, we ought to make it work. The Forest Service absolutely had an obligation to lift up Dugger and recognize it as a special place."[11]

As a result, when subcommittee hearings commenced on October 19, 1999, the only two witnesses were Congressman Riley and the Forest Service's Jack Craven, representing the Clinton administration, and they both spoke in support of the legislation.[12] "There are some parts of our environment that have disappeared because nobody thought ahead," said Congressman Riley, "but we have a unique opportunity here, right now, to save and preserve the mountain."[13]

Despite the lack of opposition, the Dugger legislation was not well received by the chairman of the full House Committee on Resources, Rep. James V. Hansen of Utah. Hansen had been a leader of antiwilderness forces in the House of Representatives for several years. His natural hostility had been exacerbated by President Clinton's totally unexpected 1996 proclamation creating the huge Grand Staircase–Escalante National Monument in southern Utah, which Hansen considered a heavy-handed interference by Washington into local affairs. The chairman was not going to let any preservation proposal go by without scrutiny.

"Congressman Riley met with Hansen and his colleagues on the committee," says former legislative aide Shana Jones. "He explained that this had been fully vented through the community and it was not Washington lawmakers trying to tell the community what was best for them, it was the community telling us what they wanted." Hansen then allowed the measure to move to the House floor.

The Dugger Mountain Wilderness Act passed the U.S. House of Representatives on a voice vote on November 1, 1999. Just the week before—an appropriate coincidence—the Clinton administration designated the Pinhoti as a "millennium legacy trail" at a White House ceremony attended by Pete Conroy and Alabama first lady Lori Siegelman. Conroy told the press, "[W]e've got mountains and more geographic diversity than any other state in the nation. . . . When people think of Alabama, they probably don't think of our natural attributes; they probably think of our cultural past. So what we're doing with this [designation] is turning the focus toward eco-tourism and environmental protection."[14]

Two days after House passage, Sen. Jeff Sessions obtained Senate approval of companion legislation, and everyone in the state thought there was nothing left but to obtain President Clinton's signature, a foregone conclusion. However, there were minor differences in the language of the House and Senate bills, which normally would have required that the measures be reconciled in a House-Senate conference. Because time was running out in the congressional session, the senator decided instead to bring Riley's bill, H.R. 2632, to the Senate floor for approval. Sessions's press assistant, John Cox, sought to calm fears that Congress would adjourn before the Senate could vote on the House bill: "Because the measure is not controversial, it's only a matter of getting it into the list of non-controversial items passed in the evening each day. We're confident that we can accomplish this, but there are a lot of items competing for approval."[15]

No one knew that one of the "competing items" would be a dairy subsidy bill. Herb Kohl of Wisconsin, a Democrat, asked Senate leadership to put a "hold" on Dugger Mountain, along with some twenty other bills, in an effort to force their sponsors to support the dairy subsidy that Kohl wanted. As it turned out, says Gerry Gilligan, Kohl got the votes he needed without involving Jeff Sessions. Thus Dugger Mountain was released from the hold and, on November 19, 1999, passed the Senate once again.

"'We opened [a bottle of wine] a few days ago when the act passed the House and Senate,' [Francine] Hutchinson relates. 'Then when it had to go back through . . . the Senate again, we just couldn't believe it. When we found out it . . . had passed [once more] I came out here [to Dugger Mountain] and shouted to the trees, "You're safe! You're safe!"'"[16]

P.L. 106-156, the Dugger Mountain Wilderness Act of 1999, was approved by President Clinton, and Alabama finally had its third national forest wilderness area—the first, ironically, to have been signed into law by a Democratic president.

Congressman Bob Riley had this to say:

I made a trip up there. After you go around and you see it's not suited for commercial harvesting of the timber, it just made sense to manage it as wilderness.

Ecologically, it is pristine. Ecotourism is going to become a bigger and bigger draw for our area—it's a win-win situation—helps the community and helps the region.

Bruce and Francine Hutchinson and Pete Conroy have been the force behind this over the years. I just want to express my appreciation for their untiring efforts for it. Without their perseverance and being a catalyst for this I'm not sure it would have been done.[17]

"We did this one right!" exclaimed Pete Conroy.

Interestingly, Dugger is just one of several federal preserves that have sprung up in northeast Alabama in recent times. Since the enactment of the Cheaha Wilderness Area in 1982, Congress has also designated the Little River Canyon National Preserve (the closest cousin to a national park that exists in Alabama) atop Lookout Mountain. Then, following the closure of the Fort McClellan Army Reservation, Congress created a 9,000-acre Mountain Longleaf National Wildlife Refuge in the Choccolocco Mountains, opening just the second publicly owned mountain range in the state, besides the Talladegas, for use by the public.

The result, says Pete Conroy, "is one of the highest concentrations of nationally protected natural areas in the country." He senses that the area's "congressional delegates, chambers of commerce, county commissions, and so forth, are now focused on and seemingly committed to ecotourism as a priority. All this in a superlative demographic that boasts 8 million people within a 120-mile radius."

Perhaps this new ethic will someday have the opportunity to be successfully tested, for there still remain publicly owned areas in the region that deserve permanent protection.

Epilogue
Whither Alabama Wilderness?

PERHAPS THE MOST REMARKABLE aspect of the Dugger Mountain legislation is that it was the only Alabama wilderness not to have been actively opposed by the U.S. Forest Service. It was also the first such proposal to have originated with the Forest Service, having been put forth by the agency as early as 1978 for its RARE II inventory. In these respects, Dugger Mountain represents a milestone and a success for the Forest Service that should be acknowledged.

Nevertheless, Dugger Mountain would never have been successfully designated if left to the Forest Service to accomplish. It required years of grassroots effort by citizen activists, including, tellingly, their challenges to Forest Service management practices that might have damaged the integrity of the area.

In any event, the question now is whether any changes in the agency suggested by the Dugger Mountain story are here to stay. Outward signs are encouraging. Witness this public exchange between Bruce Hutchinson and the Shoal Creek district ranger after passage of the Dugger Mountain legislation: "We had to make sure the Forest Service was behind this [said Hutchinson].... I can't give the Forest Service enough credit for listening to us...." [District Ranger David Merriweather replied], "Those folks are considered our partners. Our intention is to be responsible managers for the area, not as dictators but as people that can facilitate desirable management for the area."[1]

Consider also these comments by Bankhead district ranger Glen Gaines: "We're big into serving the public, that's what we're here for. I haven't had anybody come in here, asking us to go out and see more clear-cuts on the national forest. Even the logging community here knows the Bankhead is a special place. I can tell you right now we're proud as the dickens to have the Sipsey. It's a special place that we've got. It's an honor to have it."[2]

Quite a revelation, coming from representatives of a federal agency that spent more than twenty years maneuvering to defeat or severely limit Alabama wilderness initiatives. The proof of its reform, however, must come from something more substantial than statements from employees, who, after all, come and go, are promoted, fired, transferred, or retire. And as if to prove the point, just as this book was being completed, Alabama national forest supervisor Jim Gooder, the instrument of so much change in the agency here, left Alabama for a new position with a forest experimental station in Minnesota.

Thus we must look to something of more permanence, to the new generation of the Alabama forest plan that was released in draft form in 2003, at the time this book was being written. When formally adopted, the plan will govern the management of all Alabama's national forests for the next ten to fifteen years.[3]

Immediately catching the eye of the reader is this bold pronouncement: "*Timber Harvest Level Will Drop.* The 1986 Forest Plan had an allowable yearly cut of 18.5 Million Cubic Feet per year. In the revised Forest Plan the allowable cut will be 8.5 Million Cubic Feet per year. The decrease . . . is the result of a change in emphasis from balanced-age-class timber management emphasis to the health of the forest stands that are present now, and restoring forest ecosystems as a priority where the need exists."[4] The plan states that its very first goal is to "[m]anage forest and woodland ecosystems in order to restore and/or maintain native communities. . . . Emphasis will be placed on maintaining forest and plant community types not abundant on private lands."[5]

Implicit in this bureaucrat-speak is a near admission that the national forests in Alabama were mismanaged in the past. From 1930 to 1980 the emphasis of the Forest Service was to "improve forest economic yields by replacing some upland hardwood forests with faster growing loblolly pine." As perhaps the inevitable result, the "National Forests in Alabama have been experiencing a southern pine beetle epidemic since 1999 and currently more than 34,000 acres of southern yellow pine forests have been severely impacted." Especially hard hit were "densely stocked stands of loblolly" that were either promoted to occur naturally or were deliberately created by clearcutting and conversion to pine monoculture. "With large-scale mortality in these communities due to pine beetle effects, the opportunity now exists to restore these sites."[6]

The new direction, then, is to restore the national forests in Alabama to the types of woodlands that once naturally occurred in each area and can no longer be found on private land, and to promote the health of those forests so that they are less vulnerable to such epidemics as the southern pine beetle. Before leaving, Jim Gooder expressed his hope that this plan "can be a model for a lot of public land management across the nation."[7]

Bankhead district ranger Glen Gaines says that under the new plan, the guiding principle will be forest health and restoration, rather than the production of commercial timber:

> There are still uses for timber sales, but in the last [forest] plan, the idea was, we've got a volume, a target, and the district ranger would have to find places to meet that volume. Now the emphasis is on desired conditions, what tools can we use to move the forest to that condition. Timber sales can help us do that [but] whatever volume is there is not really what's pushing it.
>
> Especially in the East, national forests are a pretty limited piece of ground. They are special places. More and more people want to come here. [The Bankhead] needs to be something that private lands don't have. The remaining loblolly pine that we've got, [we will] actually restore those primarily to an upland oak-hickory forest. That's about two-thirds of the district where that emphasis will take place. And on the remaining third, in the uplands, we're looking at [restoring] fire-dependent communities that you don't find anymore in the Cumberland Plateau on private lands. We've got remnants of longleaf which are native here, as well as shortleaf which we lost a lot of [during the years timber production was emphasized].

The 2003 draft plan also contains an unprecedented number of new administrative categories for protecting special areas. Although Forest Service regulations and governing statutes have for decades mandated that the agency identify and protect areas having particular scenic, ecological, or cultural significance, in Alabama this has never before been done on any meaningful scale. Consider, for example, the 1976 Bankhead National Forest Timber Management Plan, which allocated virtually everything outside the new 12,700-acre Sipsey Wilderness and a few designated recreation and administrative sites to commercial timber production—including, incredibly, the West Fork Sipsey National Wild and Scenic River Study Area.

By contrast, the new plan effectively "zones" each forest into a wide variety of special areas and "prescribes" management activities to protect or enhance the characteristics that justified each designation. In the Bankhead, for example, there are the Kinlock, Indian Tomb Hollow and High Town Path Cultural Heritage Areas, the Flint Creek Botanical Area, and "canyon corridor" prescriptions for Caney Creek and for lower Brushy Creek and its tributaries that flow into Smith Lake. The old semiprimitive designation for Upper Thompson and Tedford Creeks has been changed to "remote backcountry recreation with few open roads," a classification that renders it "unsuitable for timber production," an improvement over the old semiprimitive prescription, which allowed some commercial logging to occur.[8] Combined with the Sipsey Wilderness and the wild and scenic river corridor, says Glen Gaines, Bankhead special designations will now total about 70,000 acres, a figure unimaginable in the days of Dick Woody and Bill Bustin. Gaines points out a particular advantage of these designations: "Old growth is one of those areas where we've got a big role to play. No one else can provide that economically [on private land]. Between the Sipsey and a lot of our canyon protections, cultural heritage, botanical areas, and backcountry around Thompson and Tedford Creek, we're looking at 70,000 acres plus that are compatible with future old growth. Plus, when you get on the uplands where a lot of the forest health and restoration activities are focused, we're going to have all the pieces that belong to this physiographic region."

In the Talladega Forest, there are backcountry designations for Oakey Mountain, for the southern flank of Dugger Mountain, and for Blue Mountain, which lies just north of Cheaha State Park. Much of the lower Talladega mountain range is prescribed for longleaf pine restoration. In south Alabama's Conecuh National Forest, there is a designated natural nonmotorized area, and even an eligible wild and scenic river along Five Runs Creek, a tributary of the Yellow River.[9] A small stretch of the Cahaba River as it flows through the Oakmulgee Division of the Talladega Forest is also declared "eligible," and the plan provides for its "outstandingly remarkable" values to be protected.[10]

And yet . . . and yet—out of the current 665,000 acres of national forest land in Alabama, the agency could identify only one suggested wilderness study area, a paltry 540-acre addition to the Cheaha designed primarily to add popular McDill Point, which was recently acquired from the state park. Areas that were semiprimitive in the 1986 plan, and should have been reclassified as backcountry, have been downgraded to a dispersed recreation prescription. In the Bankhead, the Forest Service had the opportunity to reverse the hostile whims of Howell Heflin by recommending the remainder of the West Fork Sipsey watershed for wilderness, as Congressman Flippo's initial legislation once proposed, but has chosen not to do so. Also in the Bankhead, the old

Brushy Fork RARE II area, which many, including me, believe qualifies for statutory wilderness, receives no special designation beyond "dispersed recreation," and its ravines are not even given the protection of a "canyon" prescription. For these reasons and others, the 2003 plan has been formally challenged.

The people doing the challenging are Alabama's current crop of wilderness advocates, led by an extraordinary individual named Lamar Marshall, a man who takes his wilderness seriously:

> I believe in God the Creator. In the beginning, God created the heavens and the earth. He clothed the earth with wilderness. It was born in wilderness—a cauldron of swirling ecosystems, natural processes and life-forms that make up an intricate web of interdependent living critters and processes. . . . I say without apology that people who have problems with wilderness have a problem with the Great Designer, inventor and implementer of wilderness—the Creator of Heaven and Earth. Anybody that hates wilderness hates the handiwork of God Almighty.[11]

Proudly claiming Native American, English, Welsh, and Scottish ancestry, Lamar had an upbringing typical of a great many male Alabamians. He says, "To paraphrase Hank Williams Jr., I was raised on the shotgun, skinning bucks, running trotlines, trapping fur, breaking okra, pulling corn and shelling peas. I learned to swim in the Black Warrior River."[12] Pursuing the American dream as a senior electrical designer, he helped design a nuclear power plant and was working for a large North Alabama chemical facility when he moved to an inholding in the Bankhead National Forest in 1990. There, his friend Charles Borden took him to witness firsthand the clear-cutting ways of the U.S. Forest Service. Lamar was appalled.

It didn't take long for Marshall to become motivated to act. In 1991 the Forest Service contracted a clear-cut of some 40 acres around and above a historic Native American rock shelter in a Bankhead canyon known as Indian Tomb Hollow, long considered sacred by local descendents of Cherokee inhabitants who had escaped the Great Removal of the 1830s. Outraged, Lamar patched together a photocopied bulletin, the *Bankhead Monitor*, with an article headlined "Alabama Chainsaw Massacre," attacking the Forest Service for the ravages to Indian Tomb Hollow. The *Monitor*, blunt, aggressive, and satirical, attracted immediate local support, then became a more or less regular publication focusing on Bankhead Forest preservation. It caught the attention of an anonymous donor, who offered Lamar an annual salary of twenty-four thousand dollars if he would quit his engineering job and prosecute the causes of the *Monitor* full time. He jumped at the chance. "The more I worked for polluting industries, the more I hated them and had to get out," Marshall says.[13]

In 1992 Lamar attended a conference at the University of Alabama, where he met and became fast friends with two environmental lawyers, Ray Vaughan of Montgomery, founder and director of WildLaw, and Ned Mudd of Birmingham, who is something of a folk musician besides being a practicing attorney. Assisted by these lawyers, Lamar says he "began being a perpetual plaintiff, in twenty or thirty lawsuits, against everybody from the Environmental Protection Agency to the Alabama Department of Environmental Management to the Forest Service to the U.S. Fish and Wildlife Service."[14]

In 1993 Marshall, Mudd, and Vaughan incorporated a new group, the Alabama Wilderness Alliance, to rejuvenate the cause of national forest preservation that had effectively come to an end with the Sipsey Wilderness expansion in 1988. By 1996 Lamar and company had so expanded their activities beyond the Bankhead Forest that more private funding was attracted, and the *Bankhead Monitor* evolved into a statewide organization and magazine called Wild Alabama. In this way, the axis of Wild Alabama, WildLaw, and the Alabama Wilderness Alliance has come to dominate the cause of national forest wilderness preservation in the state today. For simplicity, I'll refer to these groups collectively as the Alliance.

Folks in the Alliance have joined with national and regional groups to advocate a concept called "rewilding." Simply stated, the idea is to protect a large number of special areas in the public lands and link them together with corridors acquired from willing sellers so as to restore and promote ecological processes. In the Southeast, the Southern Appalachian Forest Coalition has developed a specific proposal to advance this vision by connecting all the Appalachian national forests, from the Bankhead and Talladega in Alabama to the Shenandoah Mountains in Virginia.[15]

The vision of the Alabama Wilderness Alliance is somewhat broader, in that it wants to protect areas throughout the state's national forests, not just in the Bankhead and Talladega. Accordingly, it has produced what it calls a phase one Alabama wilderness plan, which proposes to increase designated national forest wilderness in the state to some 264,500 acres. Such a plan, of course, could reasonably be expected to attract opposition from the timber industry and its allies in the federal Agriculture Department and in Congress. It would also require a degree of political courage and commitment by members of Alabama's congressional delegation that has never before existed. Perhaps as the Forest Service's health and restoration policies are implemented, some of this larger acreage will gain a more natural appearance and encourage a local political constituency in each area to work for its permanent protection. Meanwhile, the phase one plan appears to be more idealistic than realistic.

Nevertheless, the Alliance's ideas have led to the development of a more practical proposal having the potential of some success through the national forest planning process. Working under the auspices of the Wilderness Society, Lamar Marshall, Ken Wills, Ray Vaughan, Bruce and Francine Hutchinson, and Rickey Butch Walker have devised a variety of special designations, described in a publication titled *Alabama's Mountain Treasures*, in the Talladega and Bankhead National Forests.[16] Some of the group's ideas have, in fact, been adopted in the 2003 draft forest plan, including at least part of its proposal for enlarging the Cheaha Wilderness in the Talladega, and a Flint Creek Botanical Area and the cultural heritage areas in the Bankhead.

Since the Forest Service is now so much more responsive to citizen initiatives than in the past, perhaps the final version of the 2003 plan will, in fact, adopt some of the other *Mountain Treasures* proposals of the Wilderness Society. These include a viewshed area for the Pinhoti Trail north of Cheaha State Park, culminating in scenic areas for its route through the Shoal Creek Valley and up Augusta Mine Ridge at the northern end of the Talladega Forest. In the Bankhead, another cultural heritage designation has been suggested for historic Byler Road, as well as scenic area designation for the Caney Creek backcountry south of the Sipsey Wilderness.

But the heart of the matter is additional wilderness recommendations. *Mountain Treasures* envisions a Sipsey Wilderness addition to include upper Thompson and Tedford Creeks, supplemented by a scenic area for the Flannagin, Montgomery, and upper Borden Creek watersheds. It also recommends a new Brushy Fork Wilderness Area with national wild and scenic river designation for Brushy Fork itself, as well as for its major tributaries.

In the Talladega, it recommends wilderness designation for Oakey and Blue Mountains, and south of Cheaha State Park, a dual designation for Rebecca Mountain. Within the Hollins Wildlife Management Area, the southern part of Rebecca Mountain is proposed for scenic area designation. The northern half, running up to Alabama Highway 77, is recommended for wilderness.

Regardless of whether these proposals are ultimately endorsed by the Forest Service, they serve notice that there are still special areas to be saved in Alabama's public lands. There is recognition now within officialdom that the little national forests here are most valuable as preserves of ecosystems being rapidly degraded elsewhere. Former Alabama forest supervisor Jim Gooder worries about the impact of unchecked urban sprawl on private lands, and the fragmentation of forest ecosystems and wildlife habitat that development causes. He says:

> Public lands are absolutely critical, not only in Alabama but across the nation. In Alabama, 95 percent [of forests are] privately owned, 5 percent [are] in public lands, so you're talking about nearly 700,000 acres [of national forests in Alabama] in islands. The fragmentation issue is really [one of] wildlife, threatened and endangered species, special places that the critters need to live. There's also a need for recreation and special places like Dugger Mountain . . . that we all need to renew ourselves, to stay in touch with how special all the areas are for people. And, for the Forest Service, areas that we need to protect.

As Lamar Marshall told a reporter from the *Birmingham News*, "We've got the best forest plan in the United States, but the problem with forest plans is that they are permanently temporary."[17]

The need is for permanent protection, and the highest level allowed under the laws of the United States is that of inclusion in the National Wilderness Preservation System. It is, of course, of great benefit to obtain a wilderness recommendation from the Forest Service, if possible, as it gives the politicians cover for introducing legislation. But in truth, it is only one step toward preservation, and if history is any indicator, it may be unobtainable or even unnecessary. In the end, it is the political process that must be employed.

There is simply no substitute for solid grass-roots organizing. There should be a broad-based constituency in the congressional district for each area targeted for preservation. The effort must start with winning over the congressperson for the district, for if that individual opposes it, no senator will be willing to try to force something on the representative; nor can House colleagues from other districts or states be expected to do so.

There will always be someone to oppose wilderness—even if for no more intelligent a reason than a dislike for backpackers and birdwatchers. An effective campaign will try to anticipate and deflect the resistance, or meet it head-on. Even if the opposition is token, as, for example, with Dugger Mountain, there will still be disappointments, roadblocks, and delays. In all events, there must be a determination to devote whatever resources may be necessary and to see things through to the finish. Finally—and this is perhaps the most difficult lesson for citizen activists to accept—there must be a willingness to play the legislative game on the politicians' court.

I hope the stories of Alabama's wilderness areas will encourage those who would fight for wildness. Although some of the efforts have been daunting, a trend toward success is evident with each passing year. Politicians have generally grown more open-minded, bureaucrats more supportive, and the issue of timber production has dramatically faded in significance. But when opposition comes, perhaps our stories of wilderness battles past will give supporters a primer on how to succeed.

Most importantly, I hope that readers will understand that the last special places in Alabama's national forests are still waiting for their help.

Notes

Chapter 1

1. Statement of Mary I. Burks, hearing on S-316, Eastern Wilderness Bill, February 21, 1973.

2. I never had the pleasure of knowing Blanche Dean. The source for most of the information about her in this chapter is derived from Alice S. Christenson and L. J. Davenport, "Blanche Dean, Naturalist," *Alabama Heritage* (summer 1997).

3. Blanche E. Dean, *Trees and Shrubs of the Southeast* (Birmingham Audubon Society Press, 1988), originally published as *Trees and Shrubs in the Heart of Dixie* (Coxe Publishing, 1961, revised by Southern University Press, 1968); Dean, *Ferns of Alabama and Fern Allies* (American Southern Press, 1964); Dean, Amy Mason, Joab Thomas, *Wildflowers of Alabama and Adjoining States* (University of Alabama Press, 1973).

4. Details in this chapter relating to the formation of the Alabama Conservancy, now known as the Alabama Environmental Council, are derived from a booklet by Clara Ruth Hayman, *Protecting Alabama: A History of the Alabama Environmental Council* (Alabama Environmental Council, 1997).

5. "The Multiple-Use Sustained Yield Act of 1960," P.L. (Public Law) 86-517. Congress amended the act in 1996 by P.L. 104-3331 to ensure that "[t]he establishment and maintenance of areas of wilderness are consistent with the purposes and provisions of this Act."

6. Statement of Findings, Report of the House Committee on Agriculture, on H.R. 12025, the National Forest Timber Conservation and Management Act of 1969, 91st Congress, Report no. 91-655.

7. Burt Schorr, "Forest Service Smarts under Critics' Claim It Allowed Damage to Public Timberlands," *Wall Street Journal,* June 4, 1971.

Chapter 2

1. P.L. 88-577, the Wilderness Act, sec. 2(a).

2. Ibid., sec. 3.

3. Douglas W. Scott, "A Wilderness Forever Future—A Short History of the National Wilderness Preservation System," Campaign for America's Wilderness online report, Research Report 2001-1, June 2001, retrieved from www.leaveitwild.org.

4. Interview, David T. "Tom" Rogers, December 16, 2002.

5. Unless otherwise indicated, quotes by Bob and Mary Burks in this book are derived from an interview on July 24, 2002.

6. Alabama Forest Supervisor Del W. Thorsen, U.S. Forest Service to Mrs. Robert E. Burks Jr., President, the Alabama Conservancy, December 18, 1969.

7. Deputy Regional Forester H. C. Erickson to Mrs. Robert R. Burks Jr., December 3, 1969.

8. "Soil Management Report, Bee Branch Study Area" (National Forests in Alabama, January 1971).

9. "Field Report—Geology of the Proposed Wilderness Area in the Bankhead National Forest," Alabama Conservancy, 1970.

10. Quotes by Elberta and Robert Reid Jr. in this chapter are derived from an interview on October 21, 2002.

Chapter 3

1. This study had a variety of names. The Alabama Conservancy referred to it as the Bankhead Wilderness Study or the Sipsey Wilderness Feasibility Study. The Forest Service preferred the Bee Branch and Sipsey Fork Study Report or the Bee Branch Study. Because the Bankhead Wilderness Study was the first name it was given, I have chosen to use it in this book.
2. Tom Bevill to Mrs. Robert E. Burks Jr., December 10, 1969.
3. Robert E. Jones to Mrs. Robert E. Burks Jr., December 9, 1969.
4. Wayne J. Cloward, signing "for T. A. Schlapfer, Regional Forester," U.S. Forest Service, Atlanta, Ga., to Prof. Herbert A. McCullough, August 10, 1970.
5. Dennis M. Roth, *The Wilderness Movement and the National Forests,* 2nd ed. (Intaglio Press, 1988), 55.
6. Douglas W. Scott, "Congress's Practical Criteria for Designating Wilderness," *Wild Earth* 2, no. 1 (2001).
7. Douglas W. Scott, "A Wilderness Forever Future—A Short History of the National Wilderness Preservation System," Campaign for America's Wilderness online report, Research Report 2001-1, June 2001, retrieved from www.leaveitwild.org; emphasis in original.
8. Roth, *The Wilderness Movement and the National Forests,* 57.
9. Ibid.
10. Ibid., 16.
11. P.L. 88-577, Wilderness Act, sec. 2(c); emphasis added.
12. Unless otherwise indicated, history of the Wilderness Act and Senator Church's comments are derived from Scott, "Congress's Practical Criteria."
13. Sen. John Sparkman to Charles S. Prigmore, March 26, 1971.
14. Gov. George C. Wallace to Mr. John Orr, Forest Supervisor, March 10, 1971.
15. Undated memo from Charles Prigmore to Mary Burks, apparently sent in March 1971.
16. H.R. 8739, sec. 1 (b).
17. State Rep. Ben L. Erdreich, letter to the editor, *Birmingham News,* June 2, 1971.
18. "Conservancy Disappointed in Wilderness Bill," undated Alabama Conservancy press release.
19. "Jones-Bevill and Bee Branch," editorial, *Birmingham News,* June 1, 1971.
20. Charles Richardson, "Voice for the Wilderness May Be Blount Dixie Drawl," *Birmingham News,* October 1, 1971.
21. Rep. Bob Jones to Jean (Mrs. Leon) Tune, July 20, 1971.
22. E. W. Schultz, Deputy Chief, Forest Service, Washington, D.C., to Rep. Tom Bevill, July 2, 1971.
23. Kate Harris, "Bankhead Wilderness Area Chances 'Zero,' Official Says," *Birmingham News,* July 31, 1971.
24. James Free, "Jones and Bevill Defend Their Forest Bill," *Birmingham News,* August 5, 1971.
25. Ibid.
26. Ibid.

27. Memo, John B. Scott Jr. to Charles Prigmore and others, October 11, 1971.
28. Ibid.

Chapter 4

1. Dennis M. Roth, *The Wilderness Movement and the National Forests*, 2nd ed. (Intaglio Press, 1988), 55.

2. Mary Burks to Sen. John Sparkman, November 30, 1971.

3. Minutes of "A Meeting to Establish Criteria and Prepare a Bill Setting Up Wilderness Areas in Eastern National Forests," U.S. Forest Service, November 9, 1971.

4. Mary Burks to Sen. John Sparkman, November 30, 1971.

5. Quoted by J. Wayne Cloward, Assistant Regional Forester, Southern Region U.S. Forest Service, in address given to the Alabama Forestry Council, July 25, 1972.

6. Cloward, address given to the Alabama Forestry Council, July 25, 1972.

7. S. 3225, the Sipsey Wilderness Area Act.

8. Roth, *The Wilderness Movement and the National Forests*, 58–59.

9. Ibid., 56; Roth quotes from a letter from Sen. Herman Talmadge to Sen. Paul Fannin, October 2, 1973.

10. Peter J. Bernstein, "Conservationists Split over Bee Branch," *Birmingham News*, February 6, 1972.

11. Hearing record, S. 3224 and S. 3225, before the Subcommittee on Environment, Soil Conservation, and Forestry of the Committee on Agriculture and Forestry, U.S. Senate, July 20 and 21, 1972 (U.S. Government Printing Office, 1972), 96.

12. Arthur D. Woody, Forest Supervisor, National Forests in Alabama, to Mary Burks, June 27, 1972.

13. "Statement of Louise G. Smith, President of the Alabama Conservancy, Presented at the Hearing on Wild Areas Held by the U.S. Forest Service July 17, 1972, at Tuscaloosa, Alabama."

14. Roth, *The Wilderness Movement and the National Forests*, 59.

15. Quotes and summaries regarding the Agriculture subcommittee hearings that appear in this chapter are taken from the hearing record, S. 3224, S. 3225, and S. 3699, before the Subcommittee on Environment, Soil Conservation, and Forestry of the Committee on Agriculture and Forestry, U.S. Senate, July 20 and 21, 1972 (U.S. Government Printing Office, 1972).

16. Quotes and summaries relating to S. 3973 are taken from the report of the Senate Committee on Agriculture and Forestry, report no. 92-1214, 92nd Congress, 2nd sess., September 22, 1972.

17. Mary I. Burks, "Political History of the Sipsey Wilderness Effort and the Stand of The Alabama Conservancy on Wild Areas," presented to the Wilderness Society Conference on Southeastern Wilderness, Knoxville, Tennessee, December 2 and 3, 1972.

Chapter 5

1. Dennis M. Roth, *The Wilderness Movement and the National Forests*, 2nd ed. (Intaglio Press, 1988), 59.

2. Personal communication, Ted Snyder.

3. *Congressional Record*, vol. 119, no. 5, January 11, 1973.

4. Ibid.

5. Mary I. Burks, Wilderness Chairman, report to the quarterly meeting, board of directors, the Alabama Conservancy, January 6, 1973.

6. "Alabama Group Caught in Feud," *Huntsville Times,* January 7, 1973.

7. "Conservation Unit Neutral on Jurisdiction," *Huntsville Times,* January 9, 1973.

8. Introduced in the Senate as S. 938. Rep. Tom Bevill of Jasper became a cosponsor of the companion House legislation, H.R. 2420.

9. Statement of John R. McGuire, Chief, Forest Service, Department of Agriculture, before the Subcommittee on Public Lands of the Committee on Interior and Insular Affairs, U.S. Senate, February 21, 1973.

10. Hearing record, S. 316, Designating Certain Lands for Inclusion in the National Wilderness Preservation System, Subcommittee on Public Lands of the Committee on Interior and Insular Affairs, U.S. Senate, February 21, 1973.

11. "A Proposal, Sipsey River Eastern Wilderness Bankhead National Forest Alabama Environmental Statement," U.S. Department of Agriculture, Forest Service, March 12, 1974.

12. Sen. Herman Talmadge to Sen. Paul Fannin, October 2, 1973.

13. P.L. 90-542, the National Wild and Scenic Rivers Act, sec. 1(b).

14. Roth, *The Wilderness Movement and the National Forests,* 59.

15. Press release, Sen. George D. Aiken, May 7, 1973.

16. Report to accompany S. 3433, the Eastern Wilderness Areas Act of 1974, Committee on Agriculture and Forestry, report no. 93-803, 93d Congress, 2nd sess., May 2, 1974.

17. Sen. James B. Allen to Lyle A. Taylor, April 24, 1974.

18. Ted Bryant, "Senate Approves Wilderness Bill Protecting Sipsey, 18 Other Areas," *Birmingham Post-Herald,* June 1, 1974.

19. Roth, *The Wilderness Movement and the National Forests,* 60.

20. P.L. 93-622.

21. P.L. 93-621.

22. Personal communication, Douglas W. Scott; emphasis added.

Chapter 6

1. "Dignitaries Help Dedicate 12,000-Acre Sipsey Wilderness," *Alabama Conservation* (July–August 1975).

2. Memo, David J. Saylor to Citizens for Eastern Wilderness Members and Friends, August 14, 1974.

Chapter 7

1. John Northrop, "Highway May Send Scout Trail Down Road to Ruin," *Birmingham Post-Herald,* October 2, 1978.

2. John Northrop, "Fight Brews over Forest Road Plan," *Birmingham Post-Herald,* February 20, 1978.

3. *Birmingham News,* November 14, 1976.

4. National Environmental Policy Act, P.L. 91-190.

5. Editorial, *Birmingham News,* 1976.

Chapter 8

1. David Kepple, "Conservancy, Forestry Association Join to Find Compromise on Wilderness Issue," *Birmingham News,* December 10, 1978.

2. John Northrop, "Wilderness Area Proposal Rapped," *Birmingham Post-Herald,* October 5, 1977.

3. "Conservancy Appoints New Executive Director," press release, the Alabama Conservancy, July 7, 1977.

4. Letter, David R. Bates, Associate Director, Camping and Conservation Service, National Office—Boy Scouts of America, August 24, 1979.

5. David Hedstrom, "Forest Service Seeks Comment on Wilderness Proposal," *Centreville Press,* August 10, 1978.

6. Ibid.

7. Ibid.; Thomas F. Hill, "Foresters Seek Ideas for 16,000 Acres," *Birmingham News,* July 25, 1978.

8. Press release, Alabama Environmental Quality Association, May 15, 1979; Christina Diebold, "Scenic Drive Supporters Seeking Momentum," *Talladega Daily Home,* May 15, 1979.

9. Susan Smith, "Environmentalists' Protests May Delay Work on Drive," *Talladega Daily Home,* October 1977.

Chapter 9

1. "Wilderness Debate," *Birmingham News,* September 17, 1978.

2. P.L. 88-577, the Wilderness Act, sec. 2(a).

3. "RARE II: Gulf Coast States and Puerto Rico Supplement to Draft Environmental Statement," U.S. Forest Service, Regional Office, Atlanta, Ga., June 1978.

4. Undated letter and accompanying fact sheet from Murray Johnson in author's possession; emphasis in original.

5. Judy Johnson, "Wilderness Compromise Is Argued," *Anniston Star,* September 28, 1978.

Chapter 10

1. David Kepple, "Conservancy Not Content with U.S. Forest Service's Wilderness Area Proposals," *Birmingham News,* January 5, 1979.

2. *Birmingham News,* February 16, 1978.

3. John Northrop, "Highway May Send Scout Trail Down Road to Ruin," *Birmingham Post-Herald,* October 2, 1978; all quotes about the Odum Scout Trail in the following section are from this article.

4. Frank Blanchard, "How Much Can Nature Take before It's Not Nature?" *Montgomery Advertiser,* September 10, 1978.

5. Letter, Bill Nichols, January 29, 1979.

6. "Nichols and Scouts," editorial, *Anniston Star,* February 12, 1979.

7. Rhodes Johnston, "Scenic Trail Suit Filed," *Alabama Journal,* September 19, 1979.

8. Christina Diebold, "Scenic Drive Supporters Seeking Momentum," *Talladega Daily Home,* May 15, 1979.

9. Rhodes Johnston, "Scenic Drive EPA Statements Readied," *Alabama Journal,* May 15, 1979.

10. Press release, Alabama Environmental Quality Association, May 15, 1979.

11. Ibid.

12. Christina Diebold, "Scenic Drive Supporters Seeking Momentum," *Talladega Daily Home,* May 15, 1979.

13. Ibid.

14. Press release, Alabama Environmental Quality Association, May 15, 1979.

15. Christina Diebold, "Scenic Drive Supporters Seeking Momentum," *Talladega Daily Home,* May 15, 1979.

16. Ibid.

17. Ibid.

18. Rhodes Johnston, "Scenic Drive EPA Statements Readied," *Alabama Journal,* May 15, 1979.

19. Press release, Alabama Environmental Quality Association, May 15, 1979.

Chapter 11

1. Editorial, *Anniston Star,* date unknown; editorial, *Talladega Daily Home,* date unknown.

2. Arthur D. Woody, Forest Supervisor, National Forests in Alabama, to John N. Randolph, June 6, 1979.

3. Conversations with Bill Nichols from the author's personal recollection and file notes.

4. Conversations with Charles Kelley from the author's personal recollection and file notes.

5. David Kepple, "Scenic Drive Issue Gets New Mediator," *Birmingham News,* August 23, 1979; "Scenic Highway Dispute," editorial, *Birmingham News,* August 23, 1979.

6. Dick Forster's proposal from the author's personal recollection and file notes.

7. Stephen Barlas, "Judge May Determine Cheaha Scout Trail's Status," *Anniston Star,* August 3, 1980.

Chapter 12

1. Judy Johnson, "Wilderness Bill Shelved," *Anniston Star,* March 8, 1982.

2. Ibid.

3. "Odum Scout Trail *THREATENED AGAIN!!*" undated Alabama Conservancy mail alert.

4. "Nichols and the Trail," *Anniston Star,* March 8, 1982.

5. Tom Scarritt, "Cheaha Bill Clears Congress, Awaits Reagan's Signature," *Birmingham News,* December 21, 1982.

Chapter 13

1. David Kepple, "Conservancy, Forestry Association Join to Find Compromise on Wilderness Issue," *Birmingham News,* December 10, 1978.

2. John Northrop, "Wilderness Area Proposal Rapped," *Birmingham Post-Herald,* October 5, 1977.

3. Michael Gordon, "More Wilderness Would Hurt Timber Industries," *Florence Times-Daily,* undated file clipping in author's possession.

4. Ibid.

5. John Northrop, "Saving the Wilderness May Mean Closing the Town," *Birmingham Post-Herald,* June 24, 1978.

6. David Kepple, "Conservancy, Forestry Association Join to Find Compromise on Wilderness Issue," *Birmingham News,* December 10, 1978.

7. *Final Environmental Statement, Roadless Area Review and Evaluation* (U.S. Department of Agriculture, January 1979), F-6.

8. "Draft Regional Plan for the South" (U.S. Forest Service Southern Region, June 1981), 10.

9. *Land Suitability Analysis of the Bankhead Forest and Smith Lake Area of Alabama,* University of Alabama Regional and Urban Planning Program, Northwest Alabama Council of Local Governments, North-Central Alabama Regional Council of Governments, and Birmingham Regional Planning Commission (January 1979), 64–65.

10. "Gross Timber Income, Alabama Counties" (Alabama Cooperative Extension Service, Auburn University, 1982, 1983, 1984).

11. Harold Kennedy, "Grayson Gets 'Life-Saving' Loan," *Birmingham News,* October 8, 1980.

12. David Kepple, "Conservancy Not Content with U.S. Forest Service's Wilderness Area Proposals," *Birmingham News,* January 5, 1979.

13. John Hayman, with Clara Ruth Hayman, *A Judge in the Senate* (New South Books, 2001), 359.

Chapter 14

1. Letter, Rep. Ronnie G. Flippo, May 6, 1981.

2. Letter, Sen. Howell T. Heflin, January 16, 1979.

3. Debbie Reddin, "Borden Has a Passion for Bankhead Forest," *Decatur Daily,* May 3, 1981.

4. Unless otherwise indicated, quotes by Charles Borden in this book are derived from an interview on January 31, 2003.

5. "Forest Provides Great Get-Away," *Moulton Advertiser,* October 22, 1981.

6. "A Valuable Resource," *Decatur Daily,* December 23, 1981; "Wilderness Action Would Benefit Area," *Cullman Times,* November 26, 1981.

7. "Forest Legacy," *Florence Times-Daily,* December 2, 1981; "Nearer Home," *Huntsville Times,* October 26, 1981.

8. "Flippo and Bankhead," editorial, *Birmingham News,* March 20, 1982.

9. Letter, Sen. Howell Heflin, March 30, 1982.

10. John M. McMillan Jr. to Ronnie G. Flippo, November 6, 1981.

11. Gov. Fob James to Sen. Howell T. Heflin, March 16, 1982.

12. "Flippo Seeks to Enlarge Wilderness," *Birmingham Post-Herald,* April 2, 1982.

Chapter 15

1. Marshell Frost, letter to the editor, *Moulton Advertiser,* February 22, 1982.

2. "Gross Timber Income, Alabama Counties" (Alabama Cooperative Extension Service, Auburn University, 1982).

3. "Fact Sheet, Proposed Sipsey Wilderness Additions, H.R. 6011," U.S. Forest Service, National Forests in Alabama, Montgomery, Ala., June 25, 1982.

4. Letter, Joe J. Brown, Alabama Forest Supervisor, National Forests in Alabama, August 11, 1982; testimony of Rep. Ronnie G. Flippo before the Subcommittee on Public Lands and National Parks of the House Committee on Interior and Insular Affairs, May 24, 1982.

5. "Alabama Forest Landowner List" (Alabama Forestry Commission, 1974).

6. Judy Johnson, "Sipsey—Which Way Will Ax Fall on Forest?" *Anniston Star*, May 29, 1983.

7. "Cash Receipt Report on Forest Products Harvested in Alabama for 2001" (Alabama Forestry Commission, 2002).

8. J. L. Pate, "Heflin Seeks Compromise on Sipsey Bill," *Decatur Daily*, August 6, 1983.

9. *Proposed Revised Land and Resource Management Plan*, (National Forests in Alabama, U.S. Forest Service, 2003), 2-2.

10. Elliott Minor, "Timber Plentiful but Demand Low," Associated Press, *Birmingham Post-Herald*, September 14, 2002.

11. Bob Dunnavant, "Expanded Wilderness Foes Assailed," *Birmingham Post-Herald*, March 16, 1982.

12. Letter signed by Marshell Frost as president of the Society for the Wise Use of Federal Forest Lands and accompanying flyer and "fact sheet" distributed at a public meeting on February 22, 1982. Unless otherwise noted, all SWUFFL data in this chapter come from these materials.

13. "Fact Sheet Proposed Sipsey Wilderness Additions, H.R. 6011," U.S. Forest Service, National Forests in Alabama, Montgomery, Ala., June 25, 1982.

14. "Gross Timber Income, Alabama Counties" (Alabama Cooperative Extension Service, Auburn University, 1982).

15. "The National Forests in Alabama, 1981 Accomplishments—1982 Program of Work," U.S. Forest Service, Montgomery, Ala., February 1982.

16. Testimony of Rep. Ronnie G. Flippo before the Subcommittee on Public Lands and National Parks of the House Committee on Interior and Insular Affairs, May 24, 1982.

Chapter 16

1. Letter, Gov. Fob James, April 2, 1982.

2. Letter, Rep. Albert Lee Smith Jr., July 12, 1982.

3. Letter, Montgomery mayor Emory Folmar, August 11, 1982.

4. Bill Caton, "Special Interest Group Forces Reduction of Sipsey Acreage," *Florence Times-Daily*, May 20, 1982.

5. Bob Dunnavant, "1,000 Acres Deleted from Forest Expansion," *Birmingham Post-Herald*, May 20, 1982.

6. Letter, Sen. Howell Heflin, April 28, 1982.

7. Letter, Sen. Jeremiah Denton, April 30, 1982.

8. "Sipsey Wilderness Expansion Bill Opposed by White House," *Birmingham Post-Herald*, May 25, 1982.

9. Ibid.

10. Ed Stein, *Rocky Mountain News*, reprinted in *Birmingham Post-Herald*, May 25, 1982.

11. Testimony of Luke Slaton before the Subcommittee on Public Lands and National Parks of the House Committee on Interior and Insular Affairs, May 24, 1982.

12. "Sipsey Wilderness Expansion Bill Opposed by White House," *Birmingham Post-Herald,* May 25, 1982.

13. Gayle McCracken, "Alabama Forest Land to Be Saved Despite White House Pressure," *Birmingham Post-Herald,* August 5, 1982.

14. Letter, Sen. Howell Heflin, August 20, 1982.

15. Gayle McCracken, "Denton Request Could Kill Wilderness Bill," *Birmingham Post-Herald,* September 1982.

16. Bill Caton, "Sipsey Demonstrators Clash," *Florence Times-Daily,* September [day unknown], 1982.

17. "Protest Caravan Rolls through Shoals Saturday," *Florence Times-Daily,* September [day unknown], 1982.

18. Bill Caton, "Sipsey Demonstrators Clash," *Florence Times-Daily,* September [day unknown], 1982.

19. "Protest Caravan Rolls through Shoals Saturday," *Florence Times-Daily;* September [day unknown], 1982.

20. Letter, Sen. Jeremiah Denton, September 24, 1982.

21. "Sipsey Verdict," *Florence Times-Daily,* November 22, 1982.

22. Charles Borden, letter to the editor, *Tuscaloosa News,* September 21, 1982.

23. Tom Scarritt, "Cheaha Wilderness Idea Backed; Larger Sipsey Draws Some Opposition," *Birmingham News,* November 30, 1982.

24. Gayle McCracken, "Sipsey Wilderness Expansion Debated," *Birmingham Post-Herald,* December 1, 1982.

25. "Senate Witnesses Lock Horns over Plan That Would Triple Size of Sipsey Wilderness Area," Associated Press, *Birmingham News,* December 1, 1982.

26. Gayle McCracken, "Sipsey Wilderness Expansion Debated," *Birmingham Post-Herald,* December 1, 1982.

27. Ibid.

28. Tom Scarritt, "Cheaha Wilderness Idea Backed; Larger Sipsey Draws Some Opposition," *Birmingham News,* November 30, 1982.

Chapter 17

1. [Karl Seitz,] "Less Than Half a Loaf," *Birmingham Post-Herald,* December 22, 1982.

2. Sen. Howell Heflin to Alabama secretary of state Don Siegelman, March 10, 1983.

3. Sen. Howell Heflin to Alabama state representative Jim Bennett, April 12, 1983.

4. Record of hearing before the Subcommittee on Public Lands and National Parks of the Committee on Interior and Insular Affairs, House of Representatives, 98th Congress, April 25, 1983.

5. Gayle McCracken, "House Panel OKs Sipsey Expansion," *Birmingham Post-Herald,* May 11, 1983.

6. This quote and the following dialogue are from the hearing record before the Subcommittee on Public Lands and National Parks of the Committee on Interior and Insular Affairs, House of Representatives, 98th Congress, April 25, 1983.

7. Tom Scarritt, "Flippo's Sipsey Bill Once More Hanging on Vote of Senators," *Birmingham News,* June 7, 1983.

8. Marshell Frost, letter to the editor, *Moulton Advertiser*, June 9, 1983.

9. David J. Vann, Chairman, Water Works Board of the City of Birmingham, to Sen. Howell Heflin, May 17, 1983.

10. James May, Chairman, the Jasper Utilities Board, to Sen. Howell Heflin, June 28, 1983.

11. Quote is from Howard Adcock, President, the Smith Lake Civic Association, to Sen. Jeremiah Denton, May 11, 1983. Endorsements include: Ronny James, Chairman, the Bryan-Burnwell Water Authority, to Sen. Howell Heflin, July 27, 1983; W. D. Martin, Chairman, the City of Sumiton Water Works Board, to Sen. Howell Heflin, July 14, 1983; letter, Mayor Jerry Sadler, City of Sipsey, August 1, 1983; and Grady Perry, Chairman, Walker County Commission, to Sen. Jeremiah Denton, June 27, 1983.

12. J. L. Pate, "Heflin Seeking Compromise on Sipsey Wilderness," *Florence Times-Daily*, August 5, 1983.

13. Kelly quote from Judy Johnson, "Sipsey—Which Way Will Ax Fall on Forest?" *Anniston Star*, May 29, 1983; Miller quote from "Bankhead Report Delay Could Mean Long Denton Wait on Sipsey Decision," *Birmingham News*, July 11, 1983.

14. J. L. Pate, "Heflin Seeking Compromise Sipsey Bill," *Decatur Daily*, August 6, 1983.

15. J. L. Pate, "Heflin Seeks Compromise Sipsey Bill," *Florence Times-Daily*, August 5, 1983.

16. Ibid.

17. Ibid.

18. Testimony of J. Phil Campbell, Undersecretary of Agriculture, hearing record, S. 3224, S. 3225, and S. 3699, before the Subcommittee on Environment, Soil Conservation, and Forestry of the Committee on Agriculture and Forestry, U.S. Senate, July 20 and 21, 1972 (U.S. Government Printing Office, 1972), 3–4.

19. "Conservationists Concerned about Heflin's Sipsey Proposal," press release, Birmingham Audubon Society, August 10, 1983.

20. Laura Simmons, "Sipsey Backers Berate Heflin," *Florence Times-Daily*, August 16, 1983.

21. Les Ernst, "Heflin Told Sipsey Is Big Concern Here," *Moulton Advertiser*, August 18, 1983.

22. "About half [as large]" quote from *Birmingham News*, July 11, 1983; other information about Heflin's appearance in Birmingham from Rebecca Falkenberry to Sen. Howell Heflin, August 29, 1983.

23. Rebecca Falkenberry to Sen. Howell Heflin, August 29, 1983.

24. "Clear Choice," *Moulton Advertiser*, August 18, 1983.

25. "Sipsey 'Compromise,'" *Birmingham News*, August 25, 1983.

26. Judy Johnson, "A Matter of Life," *Anniston Star*, August 11, 1983.

27. John N. Randolph to Howell Heflin, September 29, 1983.

28. J. L. Pate, "Heflin's Sipsey Plan Wins Few Friends," *Florence Times-Daily*, September 17, 1983.

29. Randolph to Heflin, August 25, 1983.

30. Howell Heflin to John N. Randolph, October 5, 1983.

31. Randolph to Heflin, October 11, 1983.

32. Heflin to Randolph, November 1, 1983.

33. Laura Simmons, "Sipsey Bill Chances Slim during 1983," *Florence Times-Daily*, November 1, 1983.

34. "Stalling Should End," editorial, *Moulton Advertiser*, December 1, 1983.

Chapter 18

1. "Bill to Expand Sipsey Wilderness Area Stuck in Senate Panel," *Birmingham Post-Herald*, October 3, 1983.

2. *Regional Guide for the Southern Region* (U.S. Forest Service, Atlanta, Ga., 1984), 2–11.

3. *Sipsey Fork, West Fork, Draft Environmental Statement and Wild and Scenic River Study Report*, National Forests in Alabama, April 3, 1984.

4. Olivia Barton, "Sipsey River Proposal May Aid Wilderness Bill," *Birmingham News*, April 6, 1984.

5. Letter, Alice Marshall, Office of the Alabama Secretary of State, June 17, 1986.

6. Olivia Barton, "Time Waning for Senate to Pass Sipsey Expansion," *Birmingham News*, September 16, 1984.

7. Memo, Randy Snodgrass, Southeast Regional Office, Wilderness Society, Atlanta, Ga., October 10, 1984.

8. Randy Quarles, "Flippo Wants Two Senators to Make a Decision on Sipsey," *Huntsville Times*, October 7, 1984.

9. Olivia Barton, "Time Waning for Senate to Pass Sipsey Expansion," *Birmingham News*, September 16, 1984.

Chapter 19

1. John Brinkley, "Conservationists Losing Fight over Sipsey," *Birmingham Post-Herald*, February 25, 1985.

2. Viveca Novak, "Sipsey Wilderness Expansion Stirs Controversy," *Anniston Star*, February 10, 1985.

3. Olivia Barton, "Sipsey Expansion Denton, Heflin Goal," *Birmingham News*, April 23, 1985.

4. Michael Leonard to Steve Raby, Office of Sen. Howell Heflin, February 27, 1985.

5. Clint Claybrook, "Wilderness Protected While Sipsey Debated," *Decatur Daily*, April 4, 1985.

6. "Work It Out," *Decatur Daily*, April 5, 1985.

7. "Conservationists Propose 17,972-Acre Sipsey Compromise," press release, Birmingham Audubon Society, January 10, 1985.

8. John Brinkley, "Conservationists Losing Fight over Sipsey," *Birmingham Post-Herald*, February 25, 1985.

9. Clint Claybrook, "Stalemate on Sipsey Pleases Timber Industry," *Decatur Daily*, April 17, 1985.

10. Press release, Sierra Club Legal Defense Fund and Wilderness Society, April 25, 1985.

11. Ron Martz, "Environmentalists, Loggers Contest Alabama's Sipsey," *Atlanta Constitution*, May 24, 1985.

12. Letter, Sen. Jeremiah Denton, April 22, 1985.

13. John Brinkley, "Conservationists Losing Fight over Sipsey," *Birmingham Post-Herald*, February 25, 1985.

14. Randy Quarles, "Is Sipsey Wilderness Compromise in Works?" *Huntsville Times,* June 9, 1985; Olivia Barton, "Sipsey Expansion Denton, Heflin Goal," *Birmingham News,* April 23, 1985.

15. *Land and Resource Management Plan,* National Forests in Alabama, adopted October 1986, IV-67–IV-74.

16. Stan Bailey, "Groups Meet on Sipsey; Plan to Go to Senators," *Birmingham News,* July 12, 1985.

17. Susan Cullen, "Sen. Denton Aide Promises Compromise on Sipsey Soon," *Birmingham News,* July 18, 1985.

18. "Sipsey Compromise Nixed by Alabama Forestry Group," Associated Press, *Mobile Press-Register,* July 13, 1985.

19. Susan Cullen, "Sen. Denton Aide Promises Compromise on Sipsey Soon," *Birmingham News,* July 18, 1985.

20. Dale Wilson, "Groups Still at Odds over Wilderness Expansion Plan," *Montgomery Advertiser,* July 18, 1985.

21. "Denton Expects to Sponsor Bill to Expand Sipsey Wilderness," *Birmingham News,* July 30, 1985.

22. Michelle Berman, "Sipsey Wilderness Expansion Compromise Plan Unveiled," *Birmingham Post-Herald,* August 8, 1985.

23. Tom Gordon, "No Sipsey Consensus; Issue Moves to D.C.," *Birmingham News,* August 8, 1985.

24. Reactions to the meeting are from the following press reports: quotes by Richard Lee, Jim Taylor, Charlie Mitchell, and Joe Brown are from Michelle Berman, "Sipsey Wilderness Expansion Compromise Plan Unveiled," *Birmingham Post-Herald,* August 8, 1985; quotes by John Randolph, Mike Adcock, and Winston Lett are from Tom Gordon, "Latest Sipsey Plan Would Bring in Rep. Nichols," *Birmingham News,* August 11, 1985; quote by Bill Bustin is from Tom Gordon, "No Sipsey Consensus; Issue Moves to D.C.," *Birmingham News,* August 8, 1985.

25. Tom Gordon, "Latest Sipsey Plan Would Bring in Rep. Nichols," *Birmingham News,* August 11, 1985.

26. John Randolph, letter to the editor, *Birmingham News,* August 19, 1985.

27. Quotes by Mike Leonard in this book are derived from an interview on October 28, 2002.

28. The gentlemen's agreement was a letter cosigned by the Alabama Wilderness Coalition and Alabama Forestry Association that was sent to Senators Denton and Heflin and Representatives Flippo, Nichols, and Bevill in October 1985; unsigned, undated copy in author's possession.

29. Tom Gordon, "Congressional Delegation Hints at Solution to Expand Sipsey Area," *Birmingham News,* December 5, 1985.

Chapter 20

1. David Pace, "Wilderness Area Pact 90 Percent Complete," *Birmingham Post-Herald,* January 22, 1986.

2. Randy Quarles, "Sipsey Forest Agreement Seen," *Mobile Press-Register,* January 22, 1986.

3. Randy Quarles, "'Wilderness-Schmilderness' Attitude about Sipsey Troubling," *Mobile Press-Register,* January 27, 1986.

4. Michael Brumas, "Heflin Proposal Aims to End Sipsey Dispute," *Birmingham News,* May 2, 1986.

5. Howell Heflin, draft of "Sipsey Wild and Scenic River and Sipsey Wilderness Addition Act of 1986."

6. David Pace, "Wilderness Area Pact 90 Percent Complete," *Birmingham Post-Herald,* January 22, 1986.

7. Michael Brumas, "Heflin Proposal Aims to End Sipsey Dispute," *Birmingham News,* May 2, 1986.

8. Randy Quarles, "Heflin's Sipsey Proposal Could Face House Fight," *Huntsville Times,* May 2, 1986.

9. John Brinkley, "Sipsey Wilderness Dispute Smolders as Senators Stall," *Birmingham Post-Herald,* July 7, 1986.

10. Randy Quarles, "Heflin's Sipsey Proposal Could Face House Fight," *Huntsville Times,* May 2, 1986.

11. Michael Brumas, "Heflin Hopeful Sipsey Expansion Dispute Near End," *Birmingham News,* July 15, 1986.

12. Congressional Record—Senate, August 15, 1986, S. 11970-11971.

13. Press release, Sen. Howell Heflin, August 18, 1986.

Chapter 21

1. "Acreage Could Be Added to Sipsey Wilderness Area," Associated Press, *Montgomery Advertiser,* September 12, 1986.

2. Michael Brumas, "Flippo Introduces Bill to Add More Acres to Sipsey," *Birmingham News,* September 13, 1986.

3. Alabama Wilderness Coalition to Sen. Howell Heflin, May 6, 1986.

4. Michael Brumas, "Flippo Introduces Bill to Add More Acres to Sipsey," *Birmingham News,* September 13, 1986.

5. John Brinkley, "Sipsey Addition Challenged," *Birmingham Post-Herald,* September 19, 1986.

6. Robert B. McNeil, "House Subcommittee Approves Expansion Bill," *Florence Times-Daily,* September 24, 1986.

7. Except where otherwise indicated, the quotes in this segment on the Senate hearings are taken from the hearing record on S. 2782, the Sipsey Wild and Scenic River and Sipsey Wilderness Addition Act of 1986, Subcommittee on Agricultural Research, Conservation, Forestry, and General Legislation, S. Hrg. 99-1000 (U.S. Government Printing Office, 1987).

8. Dale Wilson, "Groups Still at Odds over Wilderness Expansion Plan," *Montgomery Advertiser,* June 19, 1985.

9. P.L. 88-577, the Wilderness Act, sec. 4 (d.) (1).

10. Unless otherwise indicated, quotes by Ron Tipton in this book are derived from an interview on January 29, 2003.

11. Michael Brumas, "Heflin Blasts Environmental Group for Sipsey Expansion Impasse," *Birmingham News,* October 11, 1986.

12. Bob McNeil, "Negotiations on Sipsey Stall over Beetle Control," *Florence Times-Daily*, October 11, 1986.

13. "Sipsey Wilderness Bill Again Goes Unresolved," Associated Press, *Florence Times-Daily*, October 21, 1986.

Chapter 22

1. John Brinkley, "Sipsey Deadlock Loses Dependability," *Birmingham Post-Herald*, January 19, 1987.

2. Tom Gordon, "Preservationists Retain Law Firm in Effort to Halt Expansion in Sipsey," *Birmingham News*, February 3, 1987.

3. "Sipsey Protectors Say They'll Fight in Court If Needed," *Birmingham Post-Herald*, February 4, 1987.

4. Michael Brumas, "Environmentalists Often Extremists, Heflin Says," *Birmingham News*, June 19, 1988.

5. Unless otherwise indicated, quotes by Ronnie Flippo in this book are derived from an interview on February 3, 2003.

6. Howell Heflin, draft of "Sipsey Wild and Scenic River and Sipsey Wilderness Addition Act of 1988."

7. John Brinkley, "Sipsey Wilderness Bills Gain Ground," *Birmingham Post-Herald*, September 29, 1988.

8. Unless otherwise indicated, all data about the Sipsey expansion and wild and scenic river status are from P.L. 100-547, the Sipsey Wild and Scenic River and Alabama Addition Act of 1988.

9. P.L. 90-542, the National Wild and Scenic Rivers Act, sec. 2(b).

10. David Meeks, "A Crown Jewel, Sipsey Gains Protection," *Birmingham News*, April 15, 1989.

11. Michael Brumas, "Reagan Signs Sipsey Wilderness Expansion Bill," *Birmingham News*, October 29, 1988.

12. John Brinkley, "State Environmentalists Finally Win," *Birmingham Post-Herald*, October 3, 1988.

Chapter 23

1. John Hayman, with Clara Ruth Hayman, *A Judge in the Senate* (New South Books, 2001), 349.

2. Ibid., 348–49.

Chapter 24

1. Unless otherwise indicated, details in this chapter relating to the natural and human history of Dugger Mountain are derived from "Data Summary, Dugger Mountain, Unit 1," U.S. Forest Service, National Forests in Alabama, July 1974.

2. "Potential National Natural Landmarks of the Appalachian Natural Region Ecological Report," H. R. DeSelm Department of Botany and Graduate Program in Ecology (University of Tennessee, Knoxville, for the U.S. Dept. of Interior National Park Service, 1984).

3. Angela D. Morgan, Curtis E. Hill, and Harry O. Holstein, "Archaeological Reconnaissance of Selected Portions of Dugger Mountain, Calhoun County, Alabama" (Jacksonville

State University Archaeological Resource Laboratory, Jacksonville, Ala., 1996). The "prior 'archaeological pedestrian surveys'" referred to are Curtis E. Hill and Harry O. Holstein, "Archaeological Pedestrian Survey of Selected Portions of Northeast Alabama" (Jacksonville State University, 1985, 1986, 1989, 1990).

4. Basil Penny, "Home Sweet Home," *Anniston Star,* December 5, 1999.

5. Stories of Dugger Mountain compiled by Francine Hutchinson for the Alabama Environmental Council, May 3, 1999, later published as Hutchinson, "Stories from Dugger Mountain" (Alabama Environmental Council, 1999).

6. Interview, District Ranger M. Earl Stewart, Shoal Creek Ranger District, U.S. Forest Service, April 2, 2003.

7. "Wilderness Protection for Dugger Sought," *Anniston Star,* September 1993.

8. Carlton Proctor, "Citizens Tell Views on Recreation Unit," *Anniston Star,* February 28, 1973.

9. "Data Summary, Dugger Mountain, Unit 1," 71, 80, 108.

10. "Potential National Natural Landmarks," H. R. DeSelm Department of Botany, University of Tennessee, Knoxville (September 1984).

11. Rep. Bill Nichols to Sens. Howell Heflin and Jeremiah Denton, May 1986.

12. *Land and Resource Management Plan,* National Forests in Alabama, April 1986, III-5.

13. P.L. 100-547, the Sipsey Wild and Scenic River and Alabama Addition Act of 1988.

Chapter 25

1. Unless otherwise indicated, quotes by Pete Conroy in this book are derived from an interview conducted on April 8, 2003.

2. Unless otherwise indicated, quotes by Bruce and Francine Hutchinson in this book are derived from an interview conducted on April 26, 2003.

3. Francine Hutchinson, "Stories from Dugger Mountain" (Alabama Environmental Council, 1999).

4. Ibid.

5. Justin Fox, *Birmingham News,* August 31, 1993.

6. "Data Summary, Dugger Mountain, Unit 1," National Forests in Alabama, July 1974.

7. Pat Byington to Robert Joslin, June 14, 1996.

8. Interview, David Carr, April 24, 2003.

9. Interview, Glen Browder, April 22, 2003.

10. For example, see David Pace, "Clinton Backs Dugger Project" Associated Press, *Anniston Star,* October 20, 1999, and Suzanne Struglinski, "Dugger Wilderness Act Passes House" *Anniston Star,* November 2, 1999.

11. Rep. Bill Nichols to John E. Alcock, Regional Forester, U.S. Forest Service, April 22, 1986.

Chapter 26

1. Katherine R. Dougan, "Area Wilderness Legislation Would Cap Dugger Mountain," *Anniston Star,* May 19, 1999.

2. Quotes by Gerry Gilligan in this chapter are derived from an interview on June 12, 2003.

3. Memo, David Carr to Pete Conroy and others, May 7, 1999.

4. Ibid.

5. Memo, Carla Connerly Lee of Wild Alabama (who served as secretary of the Wilderness Alliance meeting) to all participants, July 23, 1999.

6. Memo, David Carr of SELC to Wilderness Alliance meeting participants, July 30, 1999.

7. Memo, Jeff deGraffenried to Pete Conroy and others, August 1, 1999.

8. Russ Henderson, "Calling It Wild," *Anniston Star*, August 3, 1999.

9. Quotes by Shana Jones in this chapter are derived from an interview on May 23, 2003.

10. Katherine Bouma, "Strategy Shifts from Cutting to Conservation," *Birmingham News*, June 8, 2003.

11. Quotes by James A. Gooder in this chapter are derived from an interview on May 28, 2003.

12. David Pace, Associated Press, *Anniston Star*, October 20, 1999.

13. Michael Brumas, "Dugger Mountain Protection Clears Hurdle," *Birmingham News*, October 21, 1999.

14. Michael Brumas, "Pinhoti Trail Linked to Millennium Trails," *Birmingham News*, October 22, 1999.

15. Suzanne Struglinski, "Congressional Procedure, Rules Put Dugger Wilderness Act in Limbo," *Anniston Star*, October 21, 1999.

16. Katherine R. Dougan, "Go Tell It on the Mountain," *Anniston Star*, November 28, 1999.

17. Ibid.

Epilogue

1. Katherine R. Dougan, "Go Tell It on the Mountain," *Anniston Star*, November 28, 1999.

2. Quotes by District Ranger Glen Gaines in this chapter are derived from an interview on May 9, 2003.

3. Descriptions of the new forest plan are derived from the *Proposed Revised Land and Resource Management Plan, National Forests in Alabama* and the *Summary, Draft Environmental Impact Statement,* National Forests in Alabama, released in February 2003, hereafter referred to as *Proposed Plan* and *Summary EIS,* respectively.

4. *Summary EIS,* 9.

5. *Proposed Plan,* 2-9.

6. Ibid., 2-2–2-3.

7. Quotes by James A. Gooder in this chapter are derived from an interview on May 28, 2003.

8. *Proposed Plan,* 3-67.

9. Ibid., E-9.

10. *Summary EIS,* app. D, D-4.

11. Lamar Marshall, "Protecting Wilderness in the Deep South," *Journal of the National Wilderness Conference 2000* (Seattle, Wash., n.d.).

12. Ibid.

13. Nathaniel H. Axtell, "Alabama's Original 'Rednecks for Wilderness,'" *Appalachian Voice* (early winter 1999).

14. Unless otherwise indicated, quotes by Lamar Marshall are derived from an interview on May 9, 2003.

15. Hugh Irwin, Susan Andrew, and Trent Bouts, *Return the Great Forest—A Conservation Vision for the Southern Appalachian Region* (Southern Appalachian Forest Coalition, 2002).

16. *Alabama's Mountain Treasures—The Unprotected Wildlands of the Bankhead and Talladega National Forests* (Wilderness Society, 2003). Cosponsors of the report are the Alabama Environmental Council; Alabama Rivers Alliance; Alabama League of Environmental Voters; Appalachian Voices of Boone, N.C.; Bama Environmental News; Cullman Audubon Society; Dogwood Alliance of Asheville, N.C.; Environmental Policy and Information Center of Jacksonville State University; Heartwood of Bloomington, Ind.; Southern Appalachian Forest Coalition of Asheville, N.C.; Southern Environmental Law Center of Charlottesville, Va.; Tennessee Valley Audubon Society; Wild Alabama; and WildLaw.

17. Katherine Bouma, "Groups Seek to Protect 24 Forest Areas," *Birmingham News*, March 18, 2003.

Index

Acreage, inconsistencies in, xiii
Aiken, George, U.S. Senator, 28, 30, 31, 32, 33, 35, 36, 41, 42. *See also* Wild Areas
Alabama Conservancy, The, 4, 7, 9, 10, 11, 13, 17, 20, 21, 22, 26, 28, 29, 33, 37, 40, 42, 46, 55, 71, 73, 78, 80, 81, 82, 83, 99, 100, 108, 124, 181, 196; Anniston chapter of, 204; founding and early projects, 6; host of Sipsey Wilderness tour, 23; leader of Dugger Mountain Wilderness campaign, 196, 204; leader of RARE II in Alabama, 59; members' testimony before U.S. House Subcommittee on Public Lands and National Parks in support of Sipsey Wilderness Additions Act of 1983, 133; 1971 meeting with Alabama congressional delegation, 25, 30; 1972 statement on "Wild Areas" before the National Forests meeting in Alabama, 30; 1972 testimony before U.S. Senate Committee on Agriculture and Forestry on "Wild Areas" legislation, 32; participation in 1985 Congressional delegation "Gentlemen's Agreement" meetings, 161–63; participation in 1981 "Wild Areas" conference, 26–27; participation in Senator Denton's 1985 negotiations, 156–59; sponsorship of Bankhead Wilderness Feasibility Study, 11–12; statement on "Wild Areas" to 1972 meeting of The Wilderness Society, 34. *See also* Alabama Environmental Council; Talladega Scenic Drive
Alabama Cooperative Extension Service: reports of gross timber income for Alabama counties, 104, 105, 110
Alabama Department of Conservation and Natural Resources, 11, 78, 79, 86
Alabama Environmental Council, 206, 208, 214; participation in 1999 Alabama Wilderness Alliance meeting, 216–17; study of Dugger Mountain Wilderness feasibility, 196, 206–07. *See also* Alabama Conservancy
Alabama Environmental Quality Association (AEQA), 73, 74–75
Alabama Farm Bureau Federation: opposition to RARE II, 73; opposition to Sipsey Wilderness expansion, 93; use of condemnation as scare tactic, 92, 93
Alabama Forest Plan, 68, 74, 76, 79, 80, 82, 83, 84, 132, 146, 149, 151, 154, 155, 172, 202, 206; administrative designations in 2003 draft, 223–25, 227; appeal of 1985–86 Plan, 181–82, 183, 187–88, 214; forest health and restoration activities in 2003 draft of, 223–24; as justification for delay of Sipsey Wilderness expansion, 95, 122, 127, 129, 130, 133, 134, 135, 136, 141, 152; 1986 final recommendations, 166, 201, 209; timber harvest levels in 2003 draft of, 222; 2003 draft of, 211
Alabama Forestry Association (AFA), 92, 93, 131, 144, 152, 160, 168, 200; 1983 meeting of leaders with John Randolph, 143; opposition to Dugger Mountain Wilderness, 209, 211–12, 217; opposition to Sipsey Wilderness expansion, 121, 132, 149; participation in 1985 Congressional "Gentlemen's Agreement" meetings, 161–63; participation in Senator Denton's 1985 negotiations, 156–59; proposal of Sipsey Wilderness expansion, 158; rejection of "Montgomery Plan," 158
Alabama Forestry Commission: 2001 report, 105–06
Alabama Highway Department, 53, 57, 73, 76, 77, 79, 83, 84, 86, 87

Alabama's Mountain Treasures, 227
Alabama Wilderness Act of 1982, 86–87, 101–02, 124, 125
Alabama Wilderness Alliance: founding of, 214, 226; 1999 meeting of wilderness leaders of, 214–15, 226; projects of, 214, 226–27
Alabama Wilderness Coalition, 60, 63, 65 69, 70, 73, 77, 78, 80, 82, 88, 98, 99, 104, 106, 109, 136, 148, 151, 152, 167, 176, 183, 200, 215; acceptance of the "Gentlemen's Agreement," 161, 171; founders of, 59; leaders of, 83, 136; participation in 1985 Congressional delegation "Gentlemen's Agreement" meetings, 161–63; participation in Senator Denton's 1985 negotiations, 156–59; proposal of 10,000 acre reduction in Sipsey expansion, 147–48, 153, 155; rejection of Senator Heflin's National Recreation Area proposal, 145
Allen, J. Hollie, Florence Mayor: support of Sipsey Wilderness expansion, 99, 101, 123
Allen, James B. "Jim," U.S. Senator, 16, 20, 25, 37, 40, 41, 46, 52, 69, 96, 124, 213; co-sponsor of S. 1608, proposing a 12,000-acre Sipsey Wilderness Area, 21, 35; co-sponsor of S. 22, Wild Area legislation, 36; co-sponsor of S. 2216, designating West Fork Sipsey River for study as potential National Wild and Scenic River, 40; co-sponsor of S. 3224, proposing a Sipsey Wilderness Area of 6,000 acres and a Sipsey National Recreation Area of 3,000 acres, and of S. 3225, the Sipsey Wild Area Act, 28, 30, 31, 33, 35; co-sponsor of S. 3433, the Eastern Wilderness Areas Act of 1974, 42; co-sponsor of S. 3973, a National Forest Wild Areas Act, 33–34
Anniston Museum of Natural History, 203, 204, 207
Anniston Star, 66, 84, 105, 152; editorial support for Cheaha Wilderness, 71, 76, 85–86; editorial support for Dugger Mountain Wilderness, 166; editorial support for Odum Scout Trail, 71; editorial support for Sipsey Wilderness expansion, 141–42, 155

Bankhead Forest Industries, 92, 94–95
Bankhead Monitor, 225, 226
Bankhead National Forest timber management plan of 1976, 58, 223
Bankhead Wilderness Feasibility Study, 10–12; study area of, 10–11, 16, 21, 32
Bearce, Denny N., 11, 12
Bee Branch Wilderness, 21, 22, 23, 25, 45, 95, 98, 121, 124, 132, 185
Bevill, Tom, U.S. Representative, 16, 21, 23, 25, 45, 95, 98, 121, 124, 132, 185; co-sponsor of H. R. 656, proposed "Wild Areas" legislation, 36; co-sponsor of H. R. 8463, designating West Fork Sipsey River for study as potential National Wild and Scenic River, 40; co-sponsor of H. R. 8739, proposing a "Sipsey National Recreation Area" and "Bee Branch Wilderness," 21–22, 27, 35; meeting with Jim Taylor and John Randolph to discuss the "Gentlemen's Agreement," 162
Birmingham Audubon Society, 5, 6, 9, 13, 14, 45, 80, 147, 159, 186; appeal of 1985–86 Alabama Forest Plan, 181; co-founder of Alabama Wilderness Coalition, 59; establishment of Natural Area Preservation Project, 80, 82; John Randolph elected president of, 164; participation in 1985 Congressional delegation "Gentlemen's Agreement" meetings, 162–63; testimony on the Sipsey Wild and Scenic River and Wilderness Additions Act of 1986, 176
Birmingham News, 79, 172, 183, 206, 228; editorial opposition to RARE II, 63–64; editorial support for Sipsey Wilderness, 22; editorial support for Sipsey Wilderness expansion, 100, 140–41, 155; edito-

rial support for Talladega Scenic Drive, 53, 56
Birmingham Post-Herald, 6, 69, 100, 123, 153, 179, 185, 188; editorial support for Sipsey Wilderness expansion, 130–31, 155
Blue Mountain, 62, 188; draft of 2003 Alabama Forest Plan management proposal for, 224; new preservation proposal for, 227
Borden, Charles W., 109, 128, 151, 160, 180, 188, 225; background and qualifications, 98–99; Senate campaign against Howell Heflin, 146, 149, 150; support for Sipsey Wilderness expansion, 101, 106–07, 123, 139, 189
Borden Creek, 21, 25, 44, 97, 173; affected by 1986 Sipsey expansion legislation, 171; affected by proposed Sipsey Wild and Scenic River and Sipsey Wilderness Addition Act of 1986, 167; affected by the "Gentlemen's Agreement," 161; affected by the "Montgomery Plan," 156; affected by the Sipsey Wild and Scenic River and Alabama Additions Act of 1988, 185, 186; new preservation proposal for, 227; targeted for wilderness designation, 154
Borelli, Peter, 35–36, 37
Boy Scouts of America, 14, 51, 69, 70, 71, 72, 73, 74, 87, 165, 166, 199; Birmingham Council, 85; Choccolocco Council, 85; litigation by Explorer Post 15 and Troop 15, 71, 80, 88; Troop 15, Montgomery, 59
Braziel Creek, 97, 147, 173; affected by 1986 Sipsey Wilderness expansion legislation, 171; affected by proposed Sipsey Wild and Scenic River and Sipsey Wilderness Addition Act of 1986, 167; affected by the "Gentlemen's Agreement," 161; affected by the Sipsey Wild and Scenic River and Alabama Additions Act of 1988, 185; targeted for Wilderness designation, 149, 154, 157, 158

Brinkley, John, 153, 185; analysis of Sipsey Wilderness expansion controversy, 179–80
Browder, Glen, U.S. Representative, 210; as sponsor of the Dugger Mountain Wilderness Act of 1996, 208–09, 216
Brown, Joe J., Alabama National Forest Supervisor, 85, 86, 105, 132, 154, 166, 169, 186, 188; lobby for 1988 Sipsey Wilderness expansion, 181, 183, 185; "Montgomery Plan" proposal, 156; 1984 meeting with John Randolph to discuss increased recommendations for wilderness, 148–49; participation in 1985 Congressional delegation "Gentlemen's Agreement" meetings, 161–63; proposal for Dugger Mountain Wilderness Area, 158
Brushy Creek, 97, 170; affected by draft 2003 Alabama Forest Plan, 224, 225; deletion from proposed Sipsey expansion, 136, 137, 147; new preservation proposals for, 227
Brushy Fork of the Sipsey Fork of the Black Warrior River. *See* Brushy Creek
Buchanan, John H., Jr., U.S. Representative, 16–17, 32, 37; co-sponsor of H. R. 656, proposed "Wild Areas" legislation, 36; co-sponsor of H. R. 1758, Eastern Wilderness Areas Act, 36; co-sponsor of H. R. 13455, Eastern Wilderness Areas Act, 102
Burns, Pink Edward "Pinky," 197–98
Bunyan Hill Road, 10, 44; affected by the Sipsey Wild and Scenic River and Alabama Additions Act of 1988, 180, 186
Burks, Mary Ivy, x, 4, 7, 9, 10, 11, 13, 16, 17, 20, 30, 35, 37, 40, 41, 43, 46, 58, 204; co-author of Alabama Conservancy's 1971 "Wild Areas" proposal, 27, 28; education and experience, 6; 1971 meeting with Alabama congressional delegation, 25; 1972 testimony on "Wild Areas" legislation, 32; 1973 testimony before U.S. Senate Subcommittees

on Eastern Wilderness Bill, 4–5, 39; participation in 1971 "Wild Areas" conference in Washington, D.C., 26–27; praise by Senator Allen, 42; relationship with District Ranger William J. "Bill" Bustin, 45; relationship with Representative Jones, 23; relationship with Senator Aiken, 41; statement on "Wild Areas" to 1972 meeting of The Wilderness Society, Knoxville, TN, 34

Burks, Robert E. "Bob," Jr., 4, 10, 24; cofounder of The Alabama Conservancy, 6, 7; description of Charles Prigmore, 14–15

Bustin, William J., 45, 46, 103, 109, 159, 176; opposition to Sipsey Wilderness expansion, 124, 125, 132. *See also* Society for the Wise Use of Federal Forest Lands

Bylsma, James E. "Jim," 52, 62, 70, 71–72, 74, 88

Caney Creek, 224, 227

Carr, David W., Jr.: leader in Southern Appalachian Forest Coalition, 213; 1999 meeting with Representative Riley and Senator Sessions, 213–14; prosecution of Alabama Forest Plan appeal, 181–82, 183, 187–88; support of Dugger Mountain wilderness, 208, 215

Carter, Jimmy, President, 47, 58, 91, 95, 200

Casey, Ron, 100

Cheaha Mountain, 51, 52, 53, 54, 57, 69, 80, 86, 195

Cheaha State Park, 51, 54, 57, 61, 70, 73, 77, 80, 87, 188, 224, 227

Chestnutt, William F. "Bill," 132; 1983 meeting between Alabama Forestry Association leaders and John Randolph, 143

Choccolocco Mountains, 195, 220

Choccolocco Wildlife Management Area, 52, 69, 79, 86

Church, Frank, U.S. Senator, 19–20, 21, 38–39

Citizens for Eastern Wilderness, 35, 46

Civilian Conservation Corps (CCC), 57, 77, 80

Clearcutting, 99, 153, 167, 169, 198, 207, 221, 225; adoption as standard practice in National Forests, 7; as catalyst for Eastern Wilderness Movement, 7; as encouragement of pine beetle infestation, 222–23; in Indian Tomb Hollow, 225; near proposed Sipsey Wilderness, 10

Clinton, William J. "Bill," President, 203, 213; support for Dugger Mountain Wilderness, 218, 219

Condemnation, 91, 133; under the Eastern Wilderness Areas Act of 1974, 44, 64; under the Wilderness Act of 1964, 65, 187; use as scare tactic, 63, 64–66, 72, 75, 92, 93, 108, 109, 111 125

Conroy, W. Peter "Pete," 218; background and qualifications, 203–04; founder of the Anniston area chapter of the Alabama Conservancy, 204; leader of Dugger Mountain wilderness campaign, 205, 208, 210, 212, 214, 220; 1999 meeting with Forest Supervisor Jim Gooder, Senator Sessions and Representative Riley, 217; 1999 meeting with John McMillan of the Alabama Forestry Association, 211; 1999 meeting with Senator Sessions, Bruce Hutchinson and others, 211; president of the Alabama Conservancy, 204; relationship with Senator Heflin, 204; relationship with Senator Shelby, 205; relationship with U.S. Forest Service, 217

Cooper, James R. "Jim," Jr., 71, 80

Coxe, Walter F., 14

Cranal Road, 10, 44, 186

Cross Plains. *See* Piedmont

Dean, Blanche Evans, 13; educator, naturalist and environmental activist, 5, 6, 7; member of Bankhead Wilderness Feasibility Study team, 11

Decatur Daily: editorial support for Sipsey Wilderness expansion, 99, 152–53, 155

Denton, Jeremiah, U.S. Senator, 47, 99, 100, 102, 104, 109, 124, 128, 135, 138, 146, 147, 160, 161, 166, 182, 185, 201, 213; decision to sponsor Sipsey Wilderness expansion legislation, 155; defeat for reelection, 164; encounter with Senator Heflin, 160; insistence upon Alabama Forest Plan as condition for action, 122, 129, 132, 141, 149, 152; 1985 meeting with John Randolph and Jim Taylor, 162; opposition to Sipsey Wilderness expansion, 87, 126, 127, 144; proposal of Sipsey Wilderness expansion plan, 158; sponsorship of 1985 negotiating sessions for Sipsey Wilderness expansion, 156–59, 200; support for Cheaha Wilderness, 87, 126, 130; support for reduced Sipsey Wilderness expansion, 172

Dial, Gerald, State Representative, 73, 74

Dickerman, Ernest "Ernie," 9, 20; defense of Eastern wilderness, 18, 36

Dickinson, Bill, U.S. Representative, 63, 64, 65; co-sponsor of Sipsey Wilderness Additions Act of 1983, 132

Dugger Mountain: exclusion from proposed Sipsey Wild and Scenic River and Sipsey Wilderness Addition Act of 1986, 167; geology of, 195, 205; identification as potential National Natural Landmark, 200; inclusion in the "Gentlemen's Agreement," 161, 171; proposed as alternative to Cheaha Wilderness, 62, 88; proposed in lieu of larger Sipsey Wilderness expansion, 149, 158, 159, 162, 163, 200, 405; wilderness feasibility study of, 206–07

Dugger Mountain Unit, Data Summary of, 199–200, 207

Dugger Mountain Wilderness Act: of 1996, 208–09; of 1999, 216, 218, 219

Dutton, Roger, State Representative, 129, 139

Eastern Wilderness Amendments Act of 1973: as alternative to designated Eastern Wilderness, 38–40

Eastern Wilderness Areas Act: of 1973, S. 316, 36, 38, 40; of 1974, S. 3433 (enacted in 1975 and often referred to by that date), 41, 42, 44, 46, 64, 65

Eastern Wilderness Areas Act of 1975. *See* Eastern Wilderness Areas Act

Edwards, Jack, U.S. Representative: co-sponsor of Alabama Wilderness Act of 1982, 102; co-sponsor of Sipsey Wilderness Additions Act of 1983, 132

Endangered Species Act, 54, 73

Erdreich, Ben, U.S. Representative, 22, 150; co-sponsor of Sipsey Wilderness Additions Act of 1983, 132

Falkenberry, Rebecca, 88; letter to Senator Heflin, 140

Fall Creek Falls, 10, 25

Federal Highway Administration, 55, 57

Fifield, Richard "Dick," 92, 93

Flannagin Creek, 97; affected by the "Gentlemen's Agreement," 161; affected by the "Montgomery Plan," 156; affected by the Sipsey Wild and Scenic River and Alabama Additions Act of 1988, 185–86; preservation proposal for, 227

Flippo, Ronnie G., U.S. Representative, 85, 97, 99, 100, 105, 108, 123, 124, 131, 136, 137, 138, 147, 148, 154, 155, 156, 157, 159, 160, 162, 166, 167, 169, 175, 176, 178, 179, 183, 224; co-sponsor of The Sipsey Wild and Scenic River and Alabama Additions Act of 1988, 185, 188; defeat of Leo Yambreck for reelection, 127; 1985 meeting with Jim Taylor and John Randolph to discuss the "Gentlemen's Agreement," 163; 1986 meeting with Ron Tipton and John Randolph, 180–81; 1988 meeting with Mike Leonard, 183–84; overtures to Senator Heflin, 150, 172, 184, 190;

reduced acreage in Sipsey Wilderness expansion bill, 242; refusal to reintroduce Sipsey expansion legislation, 150–51; relationship with Howell Heflin, 190; sponsor of Alabama Wilderness Act of 1982, 86–87, 101–02, 112, 122, 125, 126, 127, 145; sponsor of 1986 Sipsey Expansion legislation, 171–73, 177; sponsor of Sipsey Wilderness Additions Act of 1983, 132, 133–34, 135, 139, 143

Florence Times-Daily, 121, 172; editorial support for Sipsey Wilderness expansion, 99, 127–28, 155

Folmar, Emery, Montgomery Mayor, 120–21

Forster, Richard A. "Dick," Alabama Commissioner of Conservation and Natural Resources: meeting with Alabama Wilderness Coalition leaders, 78–79; negotiation of settlement of Talladega Scenic Drive dispute, 79–80

Friends of the Earth, 26, 31

Froshin, Henry, 55, 56

Frost, Marshell, 103, 107, 108, 109, 127, 134, 152, 155, 157, 158, 180. *See also* Society for the Wise Use of Federal Forest Lands

Gaines, Glen, District Ranger, 221–22; description of draft of 2003 Alabama Forest Plan, 223, 224

"Gentlemen's Agreement," 182; provisions of, 161, 171, 188

Gibbs, George, 62, 68, 70, 72, 74, 76, 80, 81, 83, 84, 88

Gilligan, Gerry, 212, 217, 219

Gooder, James A. "Jim," Alabama National Forest Supervisor, 222; public lands management philosophy, 223, 228; support for Dugger Mountain Wilderness, 217–18

Graddick, Charles, Alabama Attorney General, 71, 101

Hagood Creek, 97, 147, 173; affected by the "Gentlemen's Agreement," 161; affected by the Sipsey Wild and Scenic River and Alabama Additions Acts of 1986 and 1988, 167, 185; proposal by U.S. Forest Service for Wilderness designation of, 149, 158

Heflin, Howell T., U.S. Senator, x, 32, 47, 96, 98, 99, 101, 104, 106, 109, 124, 128, 130, 132, 135, 136, 145, 155, 156, 163, 166, 171, 179, 180, 183, 186, 204, 209, 210, 213, 224; blame of Republicans for failure of 1982 Sipsey Wilderness expansion legislation, 126, 131; blame of the Wilderness Society for failure of his 1986 Sipsey Wilderness expansion legislation, 177–78; characterization of wilderness supporters, 138, 139, 140, 141; confrontation with angry Lawrence Countians, 139–40; confrontation with Senator Denton, 160, 162; correspondence with John Randolph regarding compromise with timber industry by, 143–44; defeat of Charles Borden for reelection, 150; insistence on Alabama Forest Plan as condition for Sipsey Wilderness expansion, 149, 165; insistence on compromise between loggers and wilderness supporters, 143–44, 145, 148, 151, 152, 153, 161, 164, 178, 200–201; insistence on pine beetle control as a condition for Sipsey Wilderness expansion, 167–68, 169–70, 173, 174, 178, 184, 185; justification of his opposition to the Sipsey Wilderness expansion, 190; misrepresentation of hunting in wilderness, 140; misrepresentation of the size of Sipsey Wilderness expansion, 140, 142; 1985 meeting with John Randolph, Jim Taylor, John McMillan and Forest Supervisor Joe Brown of, 162; sponsor of Sipsey Wild and Scenic River and Alabama Additions Act of 1988, 184; sponsor of Sipsey Wild and

Scenic River and Sipsey Wilderness Addition Act of 1986, 167–68, 169, 170, 172, 173; opposition to Sipsey Wilderness expansion, 87, 95, 100, 121, 129, 144, 180; reaction to "Gentlemen's Agreement," 161; rejection of 4,000-acre compromise reduction of Sipsey Wilderness expansion, 137; resentment of Charles Borden, 146, 149, 150, 160; support for Cheaha Wilderness, 87; Sipsey National Recreation Area proposal, 137–38, 141, 142, 144, 145, 147, 167, 170; support for original Sipsey Wilderness, 98; support for West Fork Sipsey Wild and Scenic River, 147

Herbster, Cheryl, 198, 211

Hollins Wildlife Management Area, 52, 69, 79, 227

Howell, Ruth Ann, 126, 127. *See also* Society for the Wise Use of Federal Forest Lands

Huntsville Times, 165; editorial support for Sipsey Wilderness expansion, 99–100, 155

Hutchinson, Bruce: background of, 204; coauthor of *Alabama's Mountain Treasures,* 227; leader of Dugger Mountain Wilderness campaign, 204–05, 206, 210, 216, 220; 1999 meeting with Senator Sessions, Pete Conroy and others, 212; relationship with the U.S. Forest Service, 207, 211, 217, 221

Hutchinson, Francine, 211; author of "Flora of Dugger Mountain," 205, 207; background and qualifications, 204–05; coauthor of *Alabama's Mountain Treasures,* 227; leader of Dugger Mountain Wilderness campaign, 204, 206–07, 216, 219, 220

Indian Tomb Hollow: as a Cultural Heritage Area, 224; logging in, 225

Izaak Walton League: preference for Wild Areas system, 18, 26, 29, 31; support for Eastern Wilderness, 35, 39

Jackson, Henry, U.S. Senator, 21, 29, 30, 34, 35, 36, 37, 38, 41, 43. *See also* Purity doctrine

Jacksonville State University, 198, 203, 205; faculty participation in Dugger Mountain wilderness study, 196, 206–07

James, Fob, Alabama Governor, 75, 78, 120, 132, 203; support for Sipsey Wilderness expansion, 101

Johnson, Murray, District Ranger, 62, 65, 66, 72

Jones, Robert E. "Bob," U.S. Representative, 16, 21, 25; co-sponsor of H. R. 656, proposed "Wild Areas" legislation, 36; co-sponsor of H. R. 8463, designating West Fork Sipsey River for study as National Wild and Scenic River, 40; co-sponsor of H. R. 8739, proposing a "Sipsey National Recreation Area" and "Bee Branch Wilderness," 21–22, 23, 24, 27, 32, 35; opposition to proposed Sipsey Wilderness, 21–22, 23, 36; support for Sipsey National Recreation Area in lieu of statutory wilderness, 21–22

Jones, Shana, 216, 217, 218

Kelley, Charles, Director of Game and Fish, Alabama Department of Conservation and Natural Resources: as member of Bankhead Wilderness Feasibility Study team, 11; meeting with Alabama Wilderness Coalition leaders, 78–79; support of Sipsey Wilderness expansion, 100–101

Kelley, Pete, 128, 136, 137

Kelly, Boyd: opposition to Dugger Mountain wilderness, 217; participation in Senator Denton's 1985 negotiations, 156–57

Kinlock Road, 10, 44

Kneisel, Craig, Assistant Alabama Attorney

General: assistance with Alabama Forest Plan appeal, 182, 183

Land and Resource Management Plan for the National Forests in Alabama. *See* Alabama Forest Plan
Lee, Richard, 155, 160, 161, 164, 185; conducted Senator Denton's 1985 negotiations, 156–59
Leonard, R. Michael "Mike," 84, 88, 96, 149, 151, 160, 162, 164, 203; background and qualifications, 82–83, 145–46; leader of Alabama Wilderness Coalition, 82, 83, 85, 87, 123, 136, 148, 152; negotiation of the "Gentlemen's Agreement," 161, 188; 1983 meeting with Senator Heflin, Luke Slaton and John Randolph, 137; 1987 meeting with Senator Heflin's staff, 180; 1988 meeting with Representative Flippo, 183–84; recipient of Sol Feinstone Award, 14, 82; representation of the Alabama Conservancy during Senator Denton's 1985 negotiations, 156–59; tour of Dugger Mountain with Representative Nichols and others, 166
Lett, Winston, 76, 84, 87, 159, 161, 162, 163, 166

Manasco, James, 12, 13
Manasco, Ruth, 13
Marshall, Lamar, 228; background and philosophy, 225–26; co-author of *Alabama's Mountain Treasures*, 227; environmental litigation, 226; founder of the Alabama Wilderness Alliance, 214, 226; founder of the *Bankhead Monitor*, 225; founder of Wild Alabama (now Wild South), 226
Mattox Branch, 97; affected by Representative Flippo's 1986 Sipsey Wilderness expansion legislation, 171; affected by the "Gentlemen's Agreement," 161; affected by the Sipsey Wild and Scenic River and Alabama Additions Act of 1988, 185
McAlpine, Frank, 92, 94, 95
McGuire, John R., Chief of U.S. Forest Service, 18; proposal of Eastern Wilderness Amendments Act of 1973, 38–39
McMillan, John M., Jr., 160, 168, 212; meeting with John Randolph and leaders of Alabama Forestry Association, 143; opposition to Dugger Mountain Wilderness, 211, 217; support of Sipsey Wilderness expansion as Alabama Conservation Commissioner, 101, 132; support of the "Gentlemen's Agreement" as Executive Vice President of Alabama Forestry Association, 161–63
Middleton, Rick, 155, 181, 208
Mitchell, Charles R. "Charlie": assistant to Senator Heflin, 96; assistant to Senator Allen, 41, 46; participation in reaching the "Gentlemen's Agreement," 161; representation of SWUFFL during Senator Denton's 1985 negotiations, 156–59
Montgomery Creek, 97, 153; affected by the "Gentlemen's Agreement," 161; affected by the "Montgomery Plan," 156; affected by the Sipsey Wild and Scenic River and Alabama Additions Act of 1988, 186; as dam site proposed by Senator Heflin, 167; deletion from proposed Sipsey Wilderness expansion, 147–48; new preservation proposal for, 227
"Montgomery Plan," 158, 159; description of, 156, 185
Moulton Advertiser, 99, 103, 123; editorial support for Sipsey Wilderness expansion, 140, 144, 155
Moulton–Lawrence County Chamber of Commerce: support for Sipsey Wilderness expansion, 101, 123, 129
Mountain Long Leaf National Wildlife Refuge, 195, 220
Mudd, Ned: co-founder of Alabama Wilderness Alliance, 214, 226

Index 257

Multiple use forest management, 138, 187; interpreted by timber industry, 31, 199; statutory definition of, 6

National Environmental Policy Act (NEPA), 53–54, 57, 71, 76. *See also* Talladega Scenic Drive
National Forest Land and Resource Management Plan. *See* Alabama Forest Plan
National Forest prescriptive management zones under 2003 draft of the Alabama Forest Plan, 223–25
National Forests in Alabama, 30, 45, 95, 111, 201; Alabama Forest Plan, 166; attempt to reduce size of proposed Sipsey Wilderness, 39; consent to Cheaha Wilderness, 86; efforts to defeat Cheaha Wilderness, 62, 70, 73–74, 76–77, 79, 80, 81, 84, 85, 87; efforts to defeat RARE II, 60, 61, 91; increase of designated Sipsey Wilderness Area to 12,700 acres by, 44; management of Dugger Mountain, 207–08; 1973 hearing on the Dugger Mountain Planning Unit, 198–99; 1974 Data Summary Report on the Dugger Mountain Unit, 199–200; participation in 1985 Congressional delegation "Gentlemen's Agreement" meetings, 161–63; participation in Senator Denton's 1985 negotiations, 156–59; permitted logging of Indian Tomb Hollow, 225; position on road closures in Sipsey Wilderness expansion legislation, 186; proposal of Dugger Mountain in lieu of larger Sipsey Wilderness expansion, 149, 158, 200; RARE II inventory for Alabama, 60, 61, 62, 68, 209; relocation of Odum Scout Trail, 70, 71, 72; Soil Management Report for Bankhead Wilderness Study Area, 12; study of the West Fork Sipsey River for inclusion in the National Wild and Scenic River System, 136; suspension of forestry activities in Bankhead Wilderness study area, 10; use of condemnation as scare tactic, 64–66. *See also* Alabama Forest Plan; U.S. Forest Service
National Recreation Area, 139; in lieu of Sipsey Wilderness, 21, 25, 28, 32, 33, 35, 138; in lieu of Sipsey Wilderness expansion, 141, 142, 144, 145, 147, 167, 168, 170
National Wild and Scenic Rivers Act, 40, 43, 147, 169. *See also* West Fork Sipsey River
National Wilderness Preservation System. *See* Wilderness Act of 1964
National Wildlife Federation, 39, 168
Nichols, William F. "Bill," U.S. Representative, 52, 68, 69, 70, 71, 73, 74, 76, 77, 79, 80, 81, 82, 84, 85, 88, 159, 161, 162, 208; co-sponsor of Sipsey Wild and Scenic River and Alabama Additions Act of 1988, 185; co-sponsor of Sipsey Wilderness Additions Act of 1983, 132; 1985 "Gentlemen's Agreement" meeting in Washington, 162–63; opposition to Dugger Mountain Wilderness, 166, 201, 209, 211, 215; sponsor of 1982 Cheaha Wilderness legislation, 86–87, 98, 102, 124, 125, 126, 130
Nixon, Richard M., President, 7, 24, 32, 41, 43; environmental messages to Congress, 27, 38
Northrop, John, 69, 100
Northwest Road, 10, 44; affected by proposed Sipsey Wilderness expansion legislation, 111, 139, 169, 170; affected by the Sipsey Wild and Scenic River and Alabama Additions Act of 1988, 184, 186

Oakey Mountain, 208, 216; management of under 2003 draft of the Alabama Forest Plan, 224; new preservation proposal for, 227
Odum Scout Trail, 51, 52, 53, 54, 57, 59, 60, 62, 65, 68, 69, 70, 71, 72, 74, 76, 77, 80, 82, 84, 85, 86, 87, 88, 100,

101, 130; relocation by National Forests in Alabama, 70, 71
Orr, John, Alabama National Forest Supervisor, 17, 20

Peterson, R. Max, Chief of U.S. Forest Service: opposition to Sipsey Wilderness expansion, 122, 133
Phillips, Douglas J. "Doug," 60, 123, 148
Piedmont, Alabama, 195, 196, 197, 199
Pine beetles. *See* Southern Pine Beetle
Pinhoti Trail, 70, 83, 88, 164, 216; designated as millennium legacy trail, 218
Price, James M. "Jim," 94, 148; leader of Alabama Wilderness Coalition, 83, 129, 136
Prigmore, Charles S., 8, 20, 24, 40; advocacy of compromise, 29–30, 176; background and experience, 14–15; coauthor of first Sipsey Wilderness proposal, 9; 1971 meeting with Alabama congressional delegation, 25; 1972 testimony on "Wild Areas" legislation, 32; participation in 1971 "Wild Areas" conference in Washington, 26–27; praise by Senator Allen, 42; strategy for lobbying Sipsey Wilderness, 21; testimony on the Sipsey Wild and Scenic River and Wilderness Additions Act of 1986, 176
Public water authorities: support for Sipsey Wilderness Expansion, 134–35
Purity doctrine, 29, 38, 92; as espoused by Representative Jones, 23; as espoused by U.S. Forest Service, 17–18, 24, 31, 38, 91; as refuted by members of Congress, 19–20, 36, 37

Quarles, Randy: observations on Howell Heflin, 165–66

Raby, Steve, 152
RARE II, 47, 58–59, 65, 66, 69, 73, 74, 79, 85, 91, 92, 93, 94, 97, 99, 101, 105, 111, 122, 131, 146, 148, 150, 156, 161, 171, 173, 181, 188, 190, 200, 201, 202, 215; Alabama inventory of, 60, 61, 62, 63, 64, 72, 209, 221; definition of, 129–30; opposition by Reagan Administration, 187; recommendations, 132, 133, 158
Reagan, Ronald, President, 87, 123; opposition to RARE II, 85, 94, 187; opposition to Sipsey Wilderness expansion, 94, 122, 125, 126, 131; support for Sipsey Wilderness expansion, 185
Rebecca Mountain, 218; new preservation proposal for, 227
Reid, Elberta Gibbs, 13, 17, 20, 40; description of Charles Prigmore, 15; description of Louise "Weesie" Smith, 13; description of Walter Coxe, 14; editor of "The Bankhead Forest–An Alabama Adventure," film, 14; 1971 meeting with Alabama congressional delegation, 25
Reid, Robert R. "Bob," Jr., 14, 186
Release and sufficiency language. *See* Release language
Release language, 214; definition of, 187; in Sipsey Wild and Scenic River and Alabama Additions Act of 1988, 187, 201, 215
Retan, J. Walden "Wally," 53, 54, 55, 83, 148
Riley, Robert "Bob," U.S. Representative, 210, 211, 212, 214; 1999 meeting with David Carr and Senator Sessions, 213; 1999 meeting with Pete Conroy and Forest Supervisor Jim Gooder, 217; response to timber industry concerns about Dugger Mountain, 217; sponsor of Dugger Mountain Wilderness Act of 1999, 216, 217, 218, 219–20
Road closures, 111; as proposed by Senator Heflin, 168, 169, 175; as provided in The Sipsey Wild and Scenic River and Alabama Additions Act of 1988, 186. *See also* Bunyan Hill Road; Northwest Road
Roadless Area Review and Evaluation. *See* RARE II

Robertson, F. Dale, Deputy Chief of U.S. Forest Service: opposition to Sipsey Wilderness expansion legislation, 172, 173–74, 177

Rogers, David T. "Tom": as co-author of first Sipsey Wilderness proposal, 9–10; meeting with Forest Supervisor Del W. Thorsen, 9

Roth, Dennis M., 19, 29, 34, 43

SAFC. *See* Southern Appalachian Forest Coalition

Saylor, John, U.S. Representative, 36–37, 41. *See also* Purity doctrine

Scott, Douglas W. "Doug," 18, 44

Seiberling, John, U.S. Representative, 133–34, 172–73, 176

SELC. *See* Southern Environmental Law Center

Semi-Primitive designation: applied to West Fork Sipsey watershed, 156, 161, 183, 188; definition of, 156; supplanted by draft 2003 Alabama Forest Plan, 224

Sessions, Jefferson Beauregard "Jeff," III, U.S. Senator, 210; 1999 meeting with Alabama Forestry Association officials, 217; 1999 meeting with Alabama Forest Supervisor Jim Gooder and Pete Conroy, 217; 1999 meeting with Bruce Hutchinson, Pete Conroy and others, 212; 1999 meeting with David Carr, 213–14; support of Dugger Mountain Wilderness, 212, 219

Shelby, Richard, U.S. Representative and U.S. Senator, 164, 204: co-sponsor of Sipsey Wilderness Additions Act of 1983, 132

Shoals Area Audubon Society, 190; testimony on the Sipsey Wild and Scenic River and Wilderness Additions Act of 1986, 176

Siegelman, Don, Alabama Attorney General, 101, 203; assistance with Alabama Forest Plan appeal, 182, 183, 214

Sierra Club, Alabama Chapter, 212; leader of Alabama Wilderness Coalition, 83; participation in appeal of the 1985–86 Alabama Forest Plan, 181; participation in 1985 Congressional delegation "Gentlemen's Agreement" meetings, 161–63; participation in Senator Denton's 1985 negotiations, 156–59; testimony on the Sipsey Wild and Scenic River and Wilderness Additions Act of 1986, 176

Sierra Club, The, 18, 19, 26, 30, 31, 94, 108, 129, 140, 148, 176, 214; legal defense fund, 154, 155, 181; preference for Wild Areas system, 29, 32, 35; support for Eastern Wilderness, 35–36, 39; support for Sipsey Wilderness Area, 32. *See also* Sierra Club, Alabama Chapter of; Sierra Club, Cahaba Group of

Sierra Club, Cahaba Group of, 53, 55

Sipsey Wild and Scenic River and Alabama Additions Act of 1988: provisions of, 184–87

Sipsey Wild and Scenic River and Sipsey Wilderness Addition Act of 1986: failure of, 177–78; objections to, 168–70; provisions of, 167–68; Senate hearings on, 173–76

Sipsey Wilderness expansion: impact on local timber industry, 91, 94, 95, 109–111, 140, 154; impact on state timber industry, 93, 101, 104, 106, 125, 130; opposition to by local timber industry, 104–06, 124, 127; opposition to by U.S. Forest Service, 200; public water supply protection as justification for, 134–35, 137; watershed protection as justification for, 97, 112, 137, 140, 151

Skyway Motorway, 57, 69, 70, 74, 76, 83, 87

Slaton, Luke: 1983 meeting with Senator Heflin, Mike Leonard and John Randolph, 137; support of Sipsey Wilderness Expansion, 99, 101, 123, 129. *See also Moulton Advertiser*

Smith, Albert Lee, Jr., U.S. Representative,

120, 132; co-sponsor of the Alabama Wilderness Act of 1982, 102, 129

Smith, Colonel Claude, 57, 74

Smith, Louise G. "Weesie," 26, 37; background and qualifications of, 13; co-author of Alabama Conservancy's 1971 "Wild Areas" proposal, 27, 28; member of Bankhead Wilderness Feasibility Study team, 11; 1972 testimony on "Wild Areas" legislation before U.S. Senate Subcommittee, 32; 1972 testimony on "Wild Areas" proposal by the National Forests in Alabama, 30; 1973 testimony on planning for the Dugger Mountain Unit, 199; participation in 1971 "Wild Areas" conference in Washington, 26–27; praise of by Senator Allen, 42

Smith, Mac: Cleburne County Commission President and Probate Judge, 166, 210–11

Smith, Mac J.: as Montgomery Boy Scout leader, 59, 71, 210

Snodgrass, Randy, 59, 64, 175

Snyder, Theodore "Ted," Jr., 35–36

Society for the Wise Use of Federal Forest Lands (SWUFFL), 103, 104, 106, 120, 131, 134, 143, 152, 155, 174, 176, 180; assertions of, 107–11; use of condemnation as scare tactic, 108–09, 111; opposition to Sipsey Wilderness expansion, 124, 125, 126, 127, 128, 132; participation in Senator Denton's 1985 negotiations, 156–59; participation in the "Gentlemen's Agreement," 161; proposal for an alternative Sipsey Wilderness expansion, 158

Soil: effect of intensive forestry management on, 12

Sol Feinstone Award: Alabama recipients, 14, 82

Southern Appalachian Forest Coalition (SAFC), 212, 213, 214, 226

Southern Environmental Law Center (SELC): assistance with Dugger Mountain wilderness campaign, 208, 212, 214, 215; prosecution of appeal of the 1985–86 Alabama Forest Plan, 181, 182–83

Southern Pine Beetle, 109, 179, 183; affected by the Sipsey Wild and Scenic River and Alabama Additions Act of 1988, 186–87; control of as proposed by Senator Heflin, 167–68, 169, 170, 173, 174, 176, 177, 184, 185, 222, 223; control of at Dugger Mountain, 207–08, 211, 212; control of in designated wilderness, 174, 175, 211; control of generally, 169–70

Sparkman, John J., U.S. Senator, 16, 20, 25, 26, 213; co-sponsor of S. 22, Wild Area legislation, 36; co-sponsor of S. 1608, proposing a 12,000-acre Sipsey Wilderness Area, 21, 35; co-sponsor of S. 2216, designating West Fork Sipsey River for study as potential National Wild and Scenic River, 40; co-sponsor of S. 3224, proposing a Sipsey Wilderness Area and adjoining Sipsey National Recreation Area of, and of S. 3225, the Sipsey Wild Area Act, 28, 30, 31, 33, 35; co-sponsor of S. 3433, the Eastern Wilderness Areas Act of 1974, 42; co-sponsor of S. 3973, a National Forest Wild Areas Act, 33–34

Subcommittee on Agricultural Research, Conservation, Forestry and General Legislation, of the Energy Committee of the U.S. Senate: hearings on Sipsey Wild and Scenic River and Sipsey Wilderness Addition Act of 1986 of, 173–77

Subcommittee on Public Lands and National Parks, Interior Committee of the U.S. House of Representatives, 185; hearings on Alabama Wilderness Act of 1982, 102, 122, 123–24; hearings on Representative Flippo's proposed 1986 Sipsey Wilderness expansion, 172–73; hearings on Sipsey Wilderness Additions Act of 1983, 133–34

Subcommittee on Public Lands and Reserved Water, of the Energy and Natural Resources Committee of the U.S. Senate: hearings on Sipsey Wilderness expansion of, 127, 128–29

SWUFFL. *See* Society for the Wise Use of Federal Forest Lands

TallaCoosa Highland Lakes Association, 57, 68, 74, 199

Talladega Daily Home,: editorial support for Cheaha Wilderness, 76; editorial support for Dugger Mountain Wilderness, 166

Talladega Mountains, 51, 57, 60, 71, 83, 86, 195, 220, 224

Talladega Scenic Drive, 52, 53, 57, 60, 61, 68, 69, 71, 72, 73, 74, 75, 76, 77, 79, 81, 84, 85, 87, 100–101, 166, 198; challenged by Cahaba Group of the Sierra Club, 53; compromise route negotiated by Conservation Commissioner Richard A. Forster, 80, 83; compromise route ultimately adopted, 86; early planning for, 52–53, 54; litigation against, 55–56, 71; violation of National Environmental Policy Act, 53–54

Talmadge, Herman, U.S. Senator, 28, 29, 30, 31, 32, 33, 35, 36, 40, 41

Taylor, James L. "Jim": leader of the Alabama Wilderness Coalition, 148; background and qualifications of, 147–48; participation in 1985 Congressional delegation "Gentlemen's Agreement" meetings," 161–63; representative of the Alabama Chapter of the Sierra Club during Senator Denton's 1985 negotiations, 156–59; testimony on the Sipsey Wild and Scenic River and Wilderness Additions Act of 1986, 176

Tedford Creek, 97, 147; affected by the "Gentlemen's Agreement," 161; affected by the "Montgomery Plan," 156; affected by the Sipsey Wild and Scenic River and Alabama Additions Act of 1988, 185; management under draft 2003 Alabama Forest Plan, 224; new preservation proposals for, 227; omission from Representative Flippo's 1986 Sipsey expansion legislation, 171

Thomas, Joab, 5; co-author of first Sipsey Wilderness proposal, 9

Thompson Creek, 97, 147; affected by the Sipsey Wild and Scenic River and Alabama Additions Act of 1988, 185; inclusion in Representative Flippo's 1986 Sipsey Wilderness expansion legislation, 171; inclusion in Senator Denton's Sipsey Wilderness expansion proposal, 158; inclusion in the "Gentlemen's Agreement," 161; inclusion in the "Montgomery Plan," 156; management of under draft 2003 Alabama Forest Plan, 224; new preservation proposal for, 227

Thorsen, Del W., Alabama National Forest Supervisor, 17; reaction to Sipsey Wilderness proposal, 9–10; suspension of logging in Bankhead Wilderness Study Area, 10–11

Tipton, Ronald J. "Ron," 154, 178, 184; assessment of Senator Heflin, 176–77, 183; assessment of Sipsey Wilderness expansion, 189; meeting in 1986 with Representative Flippo and John Randolph, 180–81; testimony before U.S. Senate Subcommittee on the Sipsey Wild and Scenic River and Wilderness Additions Act of 1986, 175–76

Toohey, Frank, 100, 131, 136, 138–39, 150, 184

U.S. Senate Committee on Agriculture and Forestry, 30, 40, 42, 136, 150, 152, 177; hearings on Eastern Wilderness legislation, 31–33; hearings on Wild Areas legislation, 31–33; jurisdictional dispute with U.S. Senate Committee on the Interior, 29, 31, 36, 39–40

U.S. Senate Committee on Interior and Insular Affairs, 21, 29, 31, 41, 42; hearings on Eastern wilderness legislation, 38–40. *See also* U.S. Senate Committee on Agriculture and Forestry

University of Alabama, 8, 42, 60–61, 123, 148

U.S. Department of Agriculture, 7, 8, 58, 226; opposition to a National Recreation Area for Bankhead National Forest, 32–33; opposition to proposed Sipsey Wilderness, 32–33; opposition to Sipsey Wild and Scenic River and Sipsey Wilderness Addition Act of 1986, 174–75. *See also* National Forests in Alabama; U.S. Forest Service

U.S. Forest Service, 7, 8, 10, 11, 12, 19, 20, 21, 23, 25, 30, 31, 38, 40, 41, 43, 46, 52, 53, 57, 63, 71, 73, 88, 104, 109, 110, 120, 130, 148, 151, 154, 155, 156, 161, 165, 169, 182, 186, 190, 196, 197, 198, 206, 210, 225, 226;; opposition to Eastern Wilderness, 17–18, 26, 28, 29; opposition to Sipsey Wild and Scenic River and Sipsey Wilderness Addition Act of 1986, 174–75; opposition to Sipsey Wilderness, 23, 24; opposition to Sipsey Wilderness expansion, 94, 122, 124, 129, 131, 133, 172; policy on control of the Southern Pine Beetle, 175, 176, 177, 185, 211; proposed Eastern Wilderness Amendments Act of 1973, 38–39; purity doctrine of, 17–18, 31, 37, 38, 91; report on wilderness demand in its Southern Region, 93; support for Dugger Mountain Wilderness, 217–18, 221, 228, 451, 452, 465; Wild Areas proposals, 27, 28. *See also* National Forests in Alabama; U.S. Department of Agriculture

Vaughan, Ray, 217; co-author of *Alabama's Mountain Treasures,* 227; co-founder of Alabama Wilderness Alliance, 214, 226; assistance with Alabama Forest Plan appeal, 182, 183; participation in 1999 Alabama Wilderness Alliance meeting, 215–15

Walker, Rickey Butch, 13, 14; co-author of *Alabama's Mountain Treasures,* 227

Wallace, George C., Alabama Governor, 20, 21, 22, 24, 30, 73

Water Works Board of the City of Birmingham: support for Sipsey Wilderness expansion, 134–35

Watson, J. Hilton, 132; meeting with Alabama Forestry Association leaders and John Randolph, 143

Weaver, Robert, 69–70

West Fork Sipsey River (West Fork of the Sipsey Fork of the Black Warrior River), 6, 7, 8, 9, 10, 12, 28, 47, 58, 97, 112, 151, 170, 174, 188, 223; source of public water supply, 134–35, 137, 140, 224; designation as Wild and Scenic by the Sipsey Wild and Scenic River and Alabama Additions Act of 1988, 185–86; designation for study as National Wild and Scenic River, 40, 43, 136; expansion of Wild and Scenic River corridor as catalyst for compromise Sipsey Wilderness enlargement, 147–48, 156, 158, 161; inclusion in Representative Flippo's 1986 Sipsey Wilderness expansion legislation, 171; inclusion in Senator Heflin's proposed Sipsey Wild and Scenic River and Sipsey Wilderness Addition Act of 1986, 167; inclusion in Senator Denton's proposed Sipsey Wilderness expansion, 158; inclusion in the "Gentlemen's Agreement," 161; inclusion in the "Montgomery Plan," 156; inclusion in timber industry's proposed compromise Sipsey Wilderness expansion, 158; recommendation by U.S. Forest Service for inclusion in National Wild and Scenic Rivers System, 146–47, 166, 172

Wild Alabama (now known as Wild

South), 14; participation in 1999 Alabama Wilderness Alliance meeting, 214

Wild Areas, 27, 29, 30, 31, 33, 34, 37, 38, 41, 42; advocated by Izaak Walton League, 18; advocated by Sierra Club, 18; advocated by U.S. Forest Service, 26, 27, 28; alternative to statutory wilderness, 26–27; legislation sponsored by Senator Aiken for, 28, 32, 33–34, 35, 36; legislation sponsored by Senator Talmadge for, 28, 32, 33–34, 35, 36; legislation sponsored by Senator Allen for, 28, 31–34, 35, 36; legislation sponsored by Senator Sparkman for, 28, 31–34, 35, 36

Wilderness Act of 1964, 8, 10, 17, 18, 19, 20, 28, 33, 36, 9, 42, 64, 65, 72, 175, 176, 186, 187, 228; amendment sought by Senator Heflin to permit logging to control pine beetles, 167–68, 184; amendment sought by U.S. Forest Service to establish separate criteria for Eastern areas of, 38–39; as distinguished from Eastern Wilderness Areas Act, 44; control of pine beetles pursuant to, 174–75

Wilderness, statutory: definition of, 19

Wilderness Society, The, 9, 18, 19, 20, 26, 34, 37, 43, 64, 150, 154, 168, 175, 176, 178, 183, 184, 189, 213; *Alabama's Mountain Treasures* proposals of, 227; as founder of Citizens for Eastern Wilderness, 35; as leader of Eastern Wilderness Movement, 8, 18, 29, 31, 35; as leader of RARE II effort, 59; as sponsor of 1972 Eastern Wilderness Conference in Knoxville, TN, 34; negotiations with Senator Heflin regarding control of pine beetles in Sipsey wilderness, 177; opposition to Senator Heflin's proposed 1986 Sipsey Wilderness expansion legislation, 175–76; participation in appeal of 1985 Alabama Forest Plan, 181

WildLaw, 226; participation in 1999 Alabama Wilderness Alliance meeting, 214–15

Wild South. *See* Wild Alabama

Wills, Ken, 206; co-author of *Alabama's Mountain Treasures*, 227

Woody, Arthur D. "Dick," Alabama National Forest Supervisor, 30, 44, 53, 54, 58, 72, 76–77, 83, 85, 91

Yambrek, Leo, 124; unsuccessful Republican candidate for Congress, 126, 127